Advances in Integrated Pest and Disease Management in Horticultural Crops

Advances in Integrated Pest and Disease Management in Horticultural Crops

Vol. 2

Dr. P. Singh

AGROTECH PRESS

JAIPUR ▪ NEW DELHI

AGROTECH PRESS

Regd. Off. "Bony Residency", Gate No. 2
Opp. Tilak Public School, Vishweswariya Nagar,
Gopalpura Road, Jaipur - 302018, India
Ph: 0141 - 2761280, Telefax: 2761381
E-mail : agrotechpress@gmail.com

Delhi Office: 102, 1st Floor, Satyam House,
4327/3, Ansari Road, Darya Ganj,
New Delhi - 110002, India
Ph: 011-45652440
visit us: www.agrotechpress.com

Distributed By
KEYA INTERNATIONAL PRESS LLP
E-mail: keyainternationalpressllp@gmail.com

Published in India, 2024

ISBN : 9789384568535 (4 Vols. Set)

Typeset by Thomson Press

Printed and bound in India by Thomson Press Ltd.,
New Delhi

Contents

Preface

A successful disease control program depends on a crop production system, which closely aligns with the goals of pest management. One must start with the selection of appropriate varieties, an irrigation system that minimizes leaf wetness, a fertilizer program that results in optimal plant growth, plant density and canopy management that afford optimal air circulation and pesticide coverage when needed, a transplant program that minimizes transplant shock, a clean seedling production program, effective pest monitoring by scouting regularly during the season, and, finally, a harvest and shipping procedure that maximizes shelf life and produce quality. Integrated Pest Management (IPM) as applied to vegetable diseases means using all the tactics available to the grower (cultural, biological, host-plant resistance, field scouting, chemical) that provide acceptable yield and quality at the least cost and are compatible with the tenets of environmental stewardship. IPM promotes prevention over remediation and advocates integration of multiple control strategies to achieve long-term pest management solutions.

Many insect pest and disease management problems in vegetable horticulture arise from relying entirely on pesticides for control. Integrated pest management (IPM) is a strategy that draws on a range of management tools with the goal of using the least ecologically disruptive techniques to manage pests within economically acceptable levels. Pests can include insects, mites, diseases, nematodes, weeds and vertebrate pests. IPM was first defined for managing insects and mites and in practice, most IPM growers begin with applying the strategy to insects and mites and later incorporating other pests such as diseases and nematodes. IPM covers a continuum of practices which ranges from 'intelligent pesticide management' to biointensive IPM strategies. At the most basic IPM involves routine crop monitoring, to ensure that pesticides are only applied when needed, as well as to ensure appropriate timing of pesticide applications. The most developed biointensive IPM relies primarily on beneficial organisms to manage insect pests; when greater pest control is needed, interventions chosen are complementary to the survival of these beneficials. As other pests are incorporated more and more prevention strategies are adopted which reduces the need for direct control practices. IPM is a model of continual improvement.

Vegetable growers adopting IPM typically focus initially on management strategies for a single key insect pest, usually after 'conventional' insecticides fail to control the pest adequately or are deregistered. As strategies are developed for the target pest, focus moves to other key or minor pests. IPM can become part of a fully integrated farm management system, and can potentially involve the whole market chain. IPM considers the production system in a holistic manner, and looks at all aspects of the farming enterprise as potentially increasing or decreasing pest numbers and, where possible, enhancing the activities that reduce these pest populations. In natural systems, insect pest numbers are limited by food supply, climatic conditions and natural enemies, such as predatory and parasitic insects or insect diseases. IPM seeks to enhance, rather than disrupt, this ecological balance. Where direct control is needed, a management strategy is chosen that will reduce the pest population to within economically acceptable limits with minimal adverse impact on the environment. IPM is knowledge-intensive. An IPM program is built on all compatible control tactics: cultural, biological, chemical, and mechanical; and aims over time to strengthen the preventative practices to reduce the need for control tactics. An IPM program must be an integral part of the production system as a whole, looking at the various components of the system and how they might be manipulated in order to reduce the pest population. As such, IPM could be considered as an integral component of integrated crop management (ICM), looking at all aspects necessary to grow a crop.

The book describes the scenario where integrated pest and disease control operates, that is, the vegetable horticulture site and its environment. The chapters provide the basic strategies and tactics of integrated control, with special reference to vegetable crops grown through horticulture. Further chapters include the different facets of biological pest and disease control – its scientific bases, its development in practice, its commercialization and quality control. The pre-eminence of biological control in the book is not surprising since without a doubt it is the cornerstone of integrated insect pest control and is also becoming increasingly more important in disease control. The book shows us the present situation of integrated pest and disease control in the most important vegatble crops world-wide. It also discusses the technology transfer process from research to the consumer; this aspect is by no means superfluous, as the lack of an efficient technology transfer is often the main cause of the slow adoption of integrated control.

Dr. P. Singh

1

Introduction

In culinary terms, a vegetable is an edible plant or its part, intended for cooking or eating raw. In biological terms, "vegetable" designates members of the plant kingdom.

The non-biological definition of a vegetable is largely based on culinary and cultural tradition. Apart from vegetables, other main types of plant food are fruits, grains and nuts. Vegetables are most often consumed as salads or cooked in savory or salty dishes, while culinary fruits are usually sweet and used for desserts, but it is not the universal rule. Therefore, the division is somewhat arbitrary, based on cultural views. For example, some people consider mushrooms to be vegetables even though they are not biologically plants, while others consider them a separate food category; some cultures group potatoes with cereal products such as noodles or rice, while most English speakers would consider them vegetables.

Some vegetables can be consumed raw, while some, such as cassava, must be cooked to destroy certain natural toxins or microbes in order to be edible. A number of processed food items available on the market contain vegetable ingredients and can be referred to as "vegetable derived" products. These products may or may not maintain the nutritional integrity of the vegetable used to produce them.

Examples of Different Parts of Plants used as Vegetables

The list of food items called "vegetable" is quite long, and includes many different parts of plants:

- Flower bud broccoli, cauliflower, globe artichokes, capers
- Leaves kale, collard greens, spinach, arugula, beet greens, bok choy, chard, choi sum, turnip greens, endive, lettuce, mustard greens, watercress, garlic chives, gai lan
- Leaf sheaths leeks
- Buds Brussels sprouts
- Stem Kohlrabi and galangal
- Rhizome Ginger
- Stems of leaves celery, rhubarb, cardoon, Chinese celery
- Stem shoots asparagus, bamboo shoots
- Tubers potatoes, Jerusalem artichokes, sweet potatoes, taro, and yams
- Whole-plant sprouts soybean, mung beans, urad, and alfalfa
- Roots carrots, parsnips, beets, radishes, rutabagas, turnips, and burdocks
- Bulbs onions, shallots, garlic
- Fruits in the botanical sense, but used as vegetables tomatoes, cucumbers, squash, zucchinis, pumpkins, peppers, eggplant, tomatillos, chayote, okra, breadfruit, avocado, pods, seeds such as corn, green beans and snow peas.

Nutrition

Vegetables are eaten in a variety of ways, as part of main meals and as snacks. The nutritional content of vegetables varies considerably, though generally they contain little protein or fat, and varying proportions of vitamins such as Vitamin A, Vitamin K and Vitamin B6, provitamins, dietary minerals and carbohydrates. Vegetables contain a great variety of other phytochemicals, some of which have been claimed to have antioxidant, antibacterial, antifungal, antiviral and anticarcinogenic properties. Some vegetables also contain fibre, important for gastrointestinal function. Vegetables contain important nutrients necessary for healthy hair and skin as well. A person who refrains from dairy and meat products, and eats only plants (including vegetables) is known as a vegan.

However, vegetables often also contain toxins and antinutrients such as α-solanine, α-chaconine, enzyme inhibitors (of cholinesterase, protease, amylase, etc.), cyanide and cyanide precursors, oxalic acid, and more. Depending on the concentration, such compounds may reduce the edibility, nutritional value, and health benefits of dietary vegetables. Cooking and/or other processing may be necessary to eliminate or reduce them. Diets containing recommended amounts of fruits and vegetables may help lower the risk of heart diseases and type 2 diabetes. These diets may also protect against some cancers and decrease bone loss. The potassium provided by both fruits and vegetables may help prevent the formation of kidney stones.

Dietary Recommendations

The USDA Dietary Guidelines for Americans recommends consuming 3 to 5 servings of vegetables daily. This recommendation can vary based on age and gender, and is determined based upon standard portion sizes typically consumed, as well as general nutritional content. For most vegetables, one serving is equal to 1/2 cup and can be eaten raw or cooked.

For leafy greens, such as lettuce and spinach, a single serving is typically 1 cup. Serving sizes for vegetable-derived products have not been definitively determined, but usually follow the 1/2 cup standard. Examples of vegetable-derived products subject to this standard are ketchup, pizza sauce, and tomato paste. Currently, there is no specific standard for measuring a vegetable serving in regards to its nutrient content, since different vegetables contain a wide variety of nutrients.

International dietary guidelines are similar to the ones established by the USDA. Japan, for example, recommends the consumption of 5 to 6 servings of vegetables daily. French dietary guidelines follow similar guidelines and set the daily goal at 5 servings.

Colour Pigments

The green colour of leaf vegetables is due to the presence of the green pigment chlorophyll. Chlorophyll is affected by the pH, and it changes to olive green in acid conditions, and to bright green in alkaline conditions. Some of the acids are released in steam during cooking, particularly if cooked without a cover.

The yellow/orange colours of fruits and vegetables are due to the presence of carotenoids, which are also affected by normal cooking processes or changes in pH. The red/blue colouring of some fruits and vegetables (e.g., blackberries and red cabbage) are due to anthocyanins, which are sensitive to changes in pH. When the pH is neutral, the pigments are purple, when acidic, red, and when alkaline, blue.

These pigments are quite water-soluble. This property can be used in rudimentary testing of pH.

Commercial Cultivation

Method of Vegetable Production

Most vegetables are produced using traditional farming practices. However, the yield of vegetables from organic farming is growing.

Origin Countries

Of all the world's nations, China is the leading cultivator of vegetables, with top productions in potato, onions, cabbage, lettuce, tomatoes and broccoli.

Domestic Cultivation

Amateur vegetable growing has a long history dating back to the establishment of agriculture and is a substantial business in itself, with one major online UK seed supplier currently offering well over 1200 cultivars of seeds, plants and potato tubers. They report that sale of vegetable seeds now outstrips that of flower seeds.

Each group has its own cultivation needs, but the overall requirements for the vast majority of vegetable plants are:

- deep, rich soil with a neutral or slightly alkaline composition
- plenty of sunshine
- plenty of water
- regular feeding
- regular weeding
- protection against pests such as slugs, aphids and caterpillars

The substantial literature and other media on the subject advise deep digging, annual top dressing with manure or home-made compost, and crop rotation.

Other issues are also discussed at length, including:

- raised beds
- successional sowing
- garden design
- allotments
- raising plants for competition
- comparative evaluation of different varieties
- organic gardening

Increasingly, small plants (plug plants) are offered for sale in spring and summer.

Safety

Figure: *Pulse seeds.*

For food safety, the CDC recommends proper fruit handling and preparation to reduce the risk of food contamination and foodborne illness. Fresh fruits and vegetables should be carefully selected. At the store, they should not be damaged or bruised and pre-cut pieces should be refrigerated or surrounded by ice. All fruits and vegetables should be rinsed before eating. This recommendation also applies to produce with rinds or skins that are not eaten. It should be done just before preparing or eating to avoid premature spoilage. Fruits and vegetables should be kept separate from raw foods like meat, poultry, and seafood, as well as any cooking utensils or surfaces that may have come into contact with them (e.g. cutting boards). Fruits and vegetables, if they are not going to be cooked, should be thrown away if they have touched raw meat, poultry, seafood or eggs. All cut, peeled, or cooked fruits and vegetables should be refrigerated within 2 hours. After a certain time, harmful bacteria may grow on them and increase the risk of foodborne illness.

Storage

Proper post-harvest storage aimed at extending and ensuring shelf life is best effected by efficient cold chain application. All vegetables benefit from proper post harvest care.

Many root and non-root vegetables that grow underground can be stored through winter in a root cellar or other similarly cool, dark, and dry place to prevent the growth of mold, greening and sprouting. Care should be taken in understanding the properties and vulnerabilities of the particular roots to be stored. These vegetables can last through to early spring and be nearly as nutritious as when fresh.

During storage, leafy vegetables lose moisture, and the vitamin C in them degrades rapidly. They should be stored for as short a time as possible in a cool place, in a sealed container or a plastic bag.

Figure: *"Vegetables"*

Vegetable Farming

Vegetable farming is the growing of vegetables for human consumption.

Traditionally it was done in the soil in small rows or blocks, often primarily for consumption on the farm, with the excess sold in nearby towns. Later, farms on the edge of large communities could specialize in vegetable production, with the short distance allowing the farmer to get his produce to market while still fresh. The three sisters method used by Native Americans (specifically the Haudenosaunee/Iroquois) grew squash, beans and corn together so that the plants enhanced each other's growth. Planting in long rows allows machinery to cultivate the fields, increasing efficiency and output; however, the diversity of vegetable crops requires a number of techniques to be used to optimize the growth of each type of plant. Some farms, therefore, specialize in one vegetable; others grow a large variety. Due to the needs to market vegetables while fresh, vegetable gardening has high labour demands. Some farms avoid this by running u-pick operations where the customers pick their own produce. The development of ripening technologies and refrigeration has reduced the problems with getting produce to market in good condition.

Over the past 100 years a new technique has emerged—raised bed gardening, which has increased yields from small plots of soil without the need for commercial, energy-intensive fertilizers. Modern hydroponic farming produces very high yields in greenhouses without using any soil.

Several economic models exist for vegetable farms: farms may grow large quantities of a few vegetables and sell them in bulk to major markets or middlemen, which requires large growing operations; farms may produce for local customers, which requires a larger distribution effort; farms may produce a variety of vegetables for sale through on-farm stalls, local farmer's markets, u-pick operations. This is quite different from commodity farm products like wheat and maize which do not have the ripeness problems and are sold off in bulk to the local granary. Large cities often have a central produce market which handles vegetables in a commodity-like manner, and manages distribution to most supermarkets and restaurants.

In America, vegetable farms are in some regions known as truck farms; "truck" is a noun for which its more common meaning overshadows its historically separate use as a term for "vegetables grown for market". Such farms are sometimes called muck farms, after the dark black soil in which vegetables grow well.

Common Vegetable Crops

Vegetables which are farmed include:

- Fabaceae (pea family): peas, beans, lentils
- Solanaceae (nightshade family): tomatoes, eggplants, bell peppers, potatoes
- Brassicaceae (mustard family): cauliflower, cabbage, brussels sprouts, broccoli
- Allium family: onions, garlic, leek, shallot, chives
- Carrots (Apiaceae)
- Lettuce (Asteraceae)
- cucurbit family of plants including melon, cantaloupe, cucumber, calabash, squash, and pumpkin

Post-harvest Losses (Vegetables)

The post-harvest sector includes all points in the value chain from production in the field to the food being placed on a plate for consumption. Postharvest activities include harvesting, handling, storage, processing, packaging, transportation and marketing.

Losses of horticultural produce are a major problem in the post-harvest chain. They can be caused by a wide variety of factors, ranging from growing conditions to handling at retail level. Not only are losses clearly a waste of food, but they also represent a similar waste of human effort, farm inputs, livelihoods, investments and scarce resources such as water. Post-harvest losses for horticultural produce are, however, difficult to measure. In some cases everything harvested by a farmer may end up being sold to consumers. In others, losses or waste may be considerable. Occasionally, losses may be 100%, for example when there is a price collapse and it would cost the farmer more to harvest and market the produce than to plough it back into the ground. Use of average loss figures is thus often misleading. There can be losses in quality, as measured both by the price obtained and the nutritional value, as well as in quantity.

On-farm Causes of Loss

There are numerous factors affecting post-harvest losses, from the soil in which the crop is grown to the handling of produce when it reaches the shop. Pre-harvest production practices may seriously affect post-harvest returns. Plants need a continuous supply of water for photosynthesis and transpiration. Damage can be caused by too much rain or irrigation, which can lead to decay; by too little water; and by irregular water supply, which can, for example, lead to growth cracks. Lack of plant food can affect the quality of fresh produce, causing stunted growth or discolouration of leaves, abnormal ripening and a range of other factors. Too much fertilizer can harm the development and post-harvest condition of produce. Good crop husbandry is important for reducing losses. Weeds compete with crops for nutrients and soil moisture. Decaying plant residues in the field are also a major loss factor.

Causes of Loss after Harvest

Fruits and vegetables are living parts of plant and contain 65 to 95 percent water. When food and water reserves are exhausted, produce dies and decays. Anything that increases the rate at which a product's food and water reserves are used up increases the likelihood of losses. Increases in normal physiological changes can be caused by high temperature, low atmospheric humidity and physical injury. Such injury often results from careless handling, causing internal bruising, splitting and skin breaks, thus rapidly increasing water loss.

Respiration is a continuing process in a plant and cannot be stopped without damage to the growing plant or harvested produce. It uses stored starch or sugar and stops when reserves of these are exhausted, leading to ageing. Respiration depends on a good air supply. When the air supply is restricted fermentation instead of respiration can occur. Poor ventilation of produce also leads to the accumulation of carbon dioxide. When the concentration of carbon dioxide increases it will quickly ruin produce.

Fresh produce continues to lose water after harvest. Water loss causes shrinkage and loss of weight. The rate at which water is lost varies according to the product. Leafy vegetables lose water quickly because they have a thin skin with many pores. Potatoes, on the other hand, have a thick skin with few pores. But whatever the product, to extend shelf or storage life the rate of water loss must be minimal. The most significant factor is the ratio of the surface area of the fruit or vegetable to its volume. The greater the ratio the more rapid will be the loss of water. The rate of loss is related to the difference between the water vapour pressure inside the produce and in the air. Produce must therefore be kept in a moist atmosphere.

Diseases caused by fungi and bacteria cause losses but virus diseases, common in growing crops, are not a major post-harvest problem. Deep penetration of decay makes infected produce unusable. This is often the result of infection of the produce in the field before harvest. Quality loss occurs when the disease affects only the surface. Skin blemishes may lower the sale price but do not render a fruit or vegetable inedible. Fungal and bacterial diseases are spread by microscopic spores, which are distributed in the air and soil and via decaying plant material. Infection after harvest can occur at any time. It is usually the result of harvesting or handling injuries.

Ripening occurs when a fruit is mature. Ripeness is followed by senescence and breakdown of the fruit. The category "fruit" refers also to products such as aubergine, sweet pepper and tomato. Non-climacteric fruit only ripen while still attached to the parent plant. Their eating quality suffers if they are harvested before fully ripe as their sugar and acid content does not increase further. Examples are citrus, grapes and pineapple. Early harvesting is often carried out for export shipments to minimise loss during transport, but a consequence of this is that the flavour suffers. Climacteric fruit are those that can be harvested when mature but before ripening has begun. These include banana, melon, papaya, and tomato. In commercial fruit marketing the rate of ripening is controlled artificially, thus enabling transport and distribution to be carefully planned. Ethylene gas is produced in most plant tissues and is important in starting off the ripening process. It can be used commercially for the ripening of climacteric fruits. However, natural ethylene produced by fruits can lead to in- storage losses. For example, ethylene destroys the green colour of plants. Leafy vegetables will be damaged if stored with ripening fruit. Ethylene production is increased when fruits are injured or decaying and this can cause early ripening of climacteric fruit during transport.

Damage in the Marketing Chain

Fruits and vegetables are very susceptible to mechanical injury. This can occur at any stage of the marketing chain and can result from poor harvesting practices such as the use of dirty cutting knives; unsuitable containers used at harvest time or during the marketing process, e.g. containers that can be easily squashed or have splintered wood, sharp edges or poor nailing; overpacking or underpacking of containers; and careless handling of containers. Resultant damage can include splitting of fruits, internal bruising, superficial grazing, and crushing of soft produce. Poor handling can thus result in development of entry points for moulds and bacteria, increased water loss, and an increased respiration rate.

Produce can be damaged when exposed to extremes of temperature. Levels of tolerance to low temperatures are importance when cool storage is envisaged. All produce will freeze at temperatures between 0 and -2 degrees Celsius. Although a few commodities are tolerant of slight freezing, bad temperature control in storage can lead to significant losses.

Some fruits and vegetables are also susceptible to contaminants introduced after harvest by use of contaminated field boxes; dirty water used for washing produce before packing; decaying, rejected produce lying around packing houses; and unhealthy produce contaminating healthy produce in the same packages.

Losses directly attributed to transport can be high, particularly in developing countries. Damage occurs as a result of careless handling of packed produce during loading and unloading; vibration (shaking) of the vehicle, especially on bad roads; and poor stowage, with packages often squeezed into the vehicle in order to maximise revenue for the transporters. Overheating leads to decay, and increases the rate of water loss. In transport it can result from using closed vehicles with no ventilation; stacking patterns that block the movement of air; and using vehicles that provide no protection from the sun. Breakdowns of vehicles can be a significant cause of losses in some countries, as perishable produce can be left exposed to the sun for a day or more while repairs are carried out.

At the retail marketing stage losses can be significant, particularly in poorer countries. Poor-quality markets often provide little protection for the produce against the elements, leading to rapid produce deterioration. Sorting of produce to separate the saleable from the unsaleable can result in high percentages being discarded, and there can be high weight loss from the trimming of leafy vegetables. Arrival of fresh supplies in a market may lead to some existing, older stock being discarded, or sold at very low prices.

Avoiding Loss

Losses can be avoided by following good practices as indicated above. There is also a wide range of post-harvest technologies that can be adopted to improve losses throughout the process of pre-harvest, harvest, cooling, temporary storage, transport, handling and market distribution. Recommended technologies vary depending on the type of loss experienced. However, all interventions must meet the principle of cost-effectiveness. In theory it should be possible to reduce losses substantially but in practice this may be prohibitively expensive. Especially for small farms, for which it is essential to reduce losses, it is difficult to afford expensive and work-intensive technologies.

Assessing Losses

There are no reliable methods for evaluating post-harvest losses of fresh produce. Any assessment can only refer to a particular value chain on a particular occasion and, even then, it is difficult to account for quality loss or to differentiate between unavoidable moisture loss and losses due to poor post-harvest handling and other factors described above. Accurate records of losses at various stages of the marketing chain are rarely kept, particularly in tropical countries where losses can be highest, making reliable assessment of the potential cost-effectiveness of interventions at different stages of the chain virtually impossible. The lack of such information may lead to misplaced interventions by governments and donors.

Integrated Pest Management

Integrated Pest Management (IPM) is the coordinated use of pest and environmental information to design and implement pest control methods that are economically, environmentally and socially sound. IPM promotes prevention over remediation and advocates integration of multiple control strategies to achieve long-term pest management solutions.

IPM consists of gathering information, interpreting data, creating a flexible management plan, making timely decisions and taking the proper action. Information gathering and decision-making techniques include: accurate pest identification, learning about the weak link in a pest's life-cycle or biology, scouting and monitoring crops in fields and greenhouses, using action thresholds to minimize spraying, and keeping records of findings to assess the effectiveness of management decisions.

What is IPM?

IPM stands for Integrated Pest Management. A programme started in the 19602 s IPM encourages the development of pest control methods that reduce dependence upon those insecticides that have health or environmental risks. It coordinates the use of pest biology, environmental information, and multiple pest management tactics to prevent unacceptable levels of pest damage by the most economical means.

Non-preference. (Also know as antixenosis.) Host plants that express non-preference affect the way an insect pest perceives the desireablility of the host plant. Non-preference plants either provide stimuli that are unattractive to the pest (colour, odour, texture such as downy hairs) or fail to provide stimuli that are attractive to the pest. In this way, non-preference plants affect the behaviour of pests.

Antibiosis. Antibiosis is a type of resistance in which the host plant causes injury, death, reduced longevity, or reduced reproduction of the pest. Often both a resistant and susceptible variety will have the same basic response to a pest, but the resistant variety will respond more quickly or more dramatically than the susceptible variety, reducing the amount of damage the pest causes. Plants that express antibiosis affect the biology of pests.

Tolerance. Host plants that express tolerance are resistant to pest damage because they can remain healthy and yield well despite the damage. These plants must also be able to heal wounds and fight diseases that enter through wounds.

The effectiveness of host plant resistance can vary by location because insect pests of the same species can vary somewhat by location.

Population monitoring/Economic thresholds—Monitoring pest populations and using chemical control only when pest populations reach economically damaging levels

Cultural control—Using techniques such as crop rotation to disrupt life cycles of pests that complete their life cycle in one crop

Biological control—Using natural enemy predators (e.g. lady bugs) and parasitoids (e.g. tiny stingless wasps) to suppress insect pest populations

Chemical control—Evaluating new, safer products and determining ways that all insecticides can be used more safely.

IPM Resources

Integrated crop and pest management–University of Wisconsin Extension

Vegetable IPM resource for the Midwest–University of Minnesota Extension

Pest Bulletin–Wisconsin Department of Agriculture, Trade, and Consumer Protection

Insecticide resistance: causes and action–North Central IPM Centre and Insecticide Resistance Action Committee PDF

North Central IPM Centre

Wisconsin Department of Agriculture, Trade, and Consumer Protection

Greenhouse and Hoophouse IPM

Biological control of insects and other pests of greenhouse crops–North Central Regional publication

Insect pest management for greenhouses–UW Extension publication

Hoop houses for crop extension–ATTRA webinar

Organic Resources

Vegetables–various ATTRA publications on growing organic vegetable crops

Insect control–various ATTRA publications on organic and sustainable insect management

Artichokes

The globe artichoke (*Cynara cardunculus var. scolymus*) is a variety of a species of thistle cultivated as a food. The edible portion of the plant consists of the flower buds before the flowers come into bloom. The budding artichoke flower-head is a cluster of many budding small flowers (an inflorescence) together with many bracts, on an edible base. Once the buds bloom the structure changes to a coarse, barely edible form. The uncultivated or wild variety of the species is called a cardoon. It is a perennial plant native to the Mediterranean region.

Description of the Plant

It grows to 1.4–2 m (4.6–6.6 ft) tall, with arching, deeply lobed, silvery, glaucous-green leaves 50–82 cm (20–32 in) long. The flowers develop in a large head from an edible bud about 8–15 cm (3.1–5.9 in) diametre with numerous triangular scales; the individual florets are purple. The edible portions of the buds consist primarily of the fleshy lower portions of the involucral bracts and the base, known as the "heart"; the mass of immature florets in the centre of the bud is called the "choke" or beard. These are inedible in older, larger flowers.

Early History of Use

The naturally occurring variant of the artichoke, the cardoon (*Cynara cardunculus*), which is native to the Mediterranean area, has records of use as a food among the ancient Greeks and Romans. In North Africa, where they are still found in the wild state, the seeds of artichokes, probably cultivated, were found during the excavation of Roman-period Mons Claudianus in Egypt. Varieties of artichokes were cultivated in Sicily since the classical period of the ancient Greeks, the Greeks calling them *kaktos*.

In that period the Greeks ate the leaves and flower heads, which cultivation had already improved from the wild form. The Romans called the vegetable *carduus* (whence the name cardoon). Globe artichokes are known to have been cultivated at Naples around the middle of the 9th century. Further improvement in the cultivated form appears to have taken place in the medieval period in Muslim Spain and the Maghreb, although the evidence is inferential only. Names for the artichoke in many European languages today come from medieval Arabic *al-khurshûf* via late medieval Spain.

Le Roy Ladurie, in his book *Les Paysans de Languedoc*, has documented the spread of artichoke cultivation in Italy and southern France in the late 15th and early 16th centuries, when the artichoke appears as a new arrival with a new name, which may be taken to indicate an arrival of an improved cultivated variety:

"The blossom of the thistle, improved by the Arabs, passed from Naples to Florence in 1466, carried by Filippo Strozzi. Towards 1480 it is noticed in Venice, as a curiosity. But very soon veers towards the northwest...Artichoke beds are mentioned in Avignon by the notaries from 1532 onward; from the principle [*sic*] towns they spread into the hinterlands...appearing as *carchofas* at Cavaillon in 1541, at Chateauneuf du Pape in 1553, at Orange in 1554. The local name remains *carchofas*, from the Italian *carciofo*...They are very small, the size of a hen's egg...and are still considered a luxury, a vaguely aphrodisiac tidbit that one preserved in sugar syrup."

The Dutch introduced artichokes to England, where they grew in Henry VIII's garden at Newhall in 1530. They were brought to the United States in the 19th century, to Louisiana by French immigrants and to California by Spanish immigrants.

Agricultural Output

Today, globe artichoke cultivation is concentrated in the countries bordering the Mediterranean basin. The main European producers are Italy, Spain, and France. In the United States, California provides nearly 100% of the U.S. crop, and about 80% of that is grown in Monterey County; there, Castroville proclaims itself to be "The Artichoke Centre of the World", and holds the annual Castroville Artichoke Festival. Most recently, artichokes have been grown in South Africa in a small town called Parys located along the Vaal River.

Artichokes can be produced from seeds or from vegetative means such as division, root cuttings or micropropagation. Though technically perennials that normally produce the edible flower only during the second and subsequent years, certain varieties of artichoke can be grown from seed as annuals, producing a limited harvest at the end of the first growing season, even in regions where the plants are not normally winter-hardy. This means home gardeners in northern regions can attempt to produce a crop without the need to overwinter plants with special treatment or protection. The recently introduced seed cultivar 'Imperial Star' has been bred to produce in the first year without such measures. An even newer cultivar, 'Northern Star', is said to be able to overwinter in more northerly climates, and readily survives subzero temperatures.

Commercial culture is limited to warm areas in USDA hardiness zone 7 and above. It requires good soil, regular watering and feeding, plus frost protection in winter. Rooted suckers can be planted each year, so mature specimens can be disposed of after a few years, as each individual plant lives only a few years. The peak season for artichoke harvesting is the spring, but they can continue to be harvested throughout the summer, with another peak period in midautumn. When harvested, they are cut from the plant so as to leave an inch or two of stem. Artichokes possess good keeping qualities, frequently remaining quite fresh for two weeks or longer under average retail conditions.

Apart from food use, the globe artichoke is also an attractive plant for its bright floral display, sometimes grown in herbaceous borders for its bold foliage and large purple flower heads.

Varieties

Figure: *Some varieties of artichoke display purple colouration*

Traditional Cultivars (vegetative propagation)

- Green, big: 'Camus de Bretagne', 'Castel' (France), 'Green Globe' (USA, South Africa)
- *Green, medium-size:* 'Verde Palermo' (Sicily), 'Blanca de Tudela' (Spain), 'Argentina', 'Española' (Chile), 'Blanc d'Oran' (Algeria), 'Sakiz', 'Bayrampasha' (Turkey)
- *Purple, big:* 'Romanesco', 'C3' (Italy)
- *Purple, medium-size:* 'Violet de Provence' (France), 'Brindisino', 'Catanese', 'Niscemese' (Sicily), 'Violet d'Algerie' (Algeria), 'Baladi' (Egypt), 'Ñato' (Argentina), 'Violetta di Chioggia' (Italy)
- *Spined:* 'Spinoso Sardo e Ingauno' (Sardinia), 'Criolla' (Peru).

Cultivars Propagated by Seeds

- *For industry:* 'Madrigal', 'Lorca', 'A-106', 'Imperial Star'
- *Green:* 'Symphony', 'Harmony'
- *Purple:* 'Concerto', 'Opal', 'Tempo'

Uses

Cooking: In the US, large globe artichokes are frequently prepared by removing all but 5–10 mm (0.2–0.4 in) or so of the stem. To remove thorns, which may interfere with eating, around a quarter of each scale can be cut off. To cook, the artichoke is boiled or steamed. The core of the stem tastes similar to the artichoke heart, and is edible.

Salt may be added to the water if boiling artichokes. Leaving the pot uncovered may allow acids to boil off. Covered artichokes, in particular those that have been cut, can turn brown due to the enzymatic browning and chlorophyll oxidation. Placing them in water slightly acidified with vinegar or lemon juice can prevent the discolouration.

Leaves are often removed one at a time, and the fleshy base eaten, with hollandaise, vinegar, butter, mayonnaise, aioli, lemon juice, or other sauces. The fibrous upper part of each leaf is usually discarded. The heart is eaten when the inedible choke has been peeled away from the base and discarded. The thin leaves covering the choke are also edible.

In Italy, artichoke hearts in oil are the usual vegetable for 'spring' section of the 'Four Seasons' pizza (with olives for summer, mushrooms for autumn, and prosciutto for winter). A recipe well known in Rome is Jewish-style artichokes, which are deep-fried whole.

Stuffed artichoke recipes are abundant. A common Italian stuffing uses a mixture of bread crumbs, garlic, oregano, parsley, grated cheese, and prosciutto or sausage. A bit of the mixture is then pushed into the spaces at the base of each leaf and into the centre before boiling or steaming. A similar recipe is popular in coastal Croatia.

In Spain, the more tender, younger, and smaller artichokes are used. They can be sprinkled with olive oil and left in hot ashes in a barbecue, sauteed in olive oil with garlic, with rice as a paella, or sautéed and combined with eggs in a tortilla (frittata).

Often cited is the Greek, *aginares a la polita* (artichokes city-style, referring to the city of Constantinople), a hearty, savory stew made with artichoke hearts, potatoes, and carrots, and flavoured with onion, lemon, and dill. The finest examples are to be found on the island of Tinos, and in Iria and Kantia, two small villages in Argolida in the Peloponnese of southern Greece.

Another way to use artichokes is to completely break off all of the leaves, leaving the bare heart. The leaves are steamed to soften the fleshy base part of each leaf to be used as the basis for any number of side dishes or appetizing dips, or the fleshy part is left attached to the heart, while the upper parts of the leaves are discarded. The remaining concave-shaped heart is often filled with meat, then fried or baked in a savory sauce. Frozen artichoke hearts are a time-saving substitute, though the consistency and stronger flavour of fresh hearts when available is preferred.

Throughout North Africa, the Middle East, Turkey, and Armenia, a favourite filling for stuffed artichoke hearts includes ground lamb. Spices reflect the local cuisine of each country. In Lebanon, for example, the typical filling would include lamb, onion, tomato, pinenuts, raisins, parsley, dill, mint, black pepper, and allspice. A popular Turkish vegetarian variety uses only onion, carrot, green peas, and salt.

Herbal Tea

Artichokes can also be made into an herbal tea. "Artichoke tea" is produced as a commercial product in the Da Lat region of Vietnam. The flower portion is put into water and consumed as an herbal tea, called *alcachofa* in Mexico. It has a slightly bitter woody taste.

Liqueur

Artichoke is the primary flavour of the 33-proof (16.5%-alcohol) Italian liqueur Cynar produced exclusively by the Campari Group. It can be served over ice as an aperitif or as a cocktail mixed with orange juice, especially popular in Switzerland. It is also used to make a 'Cin Cyn', a slightly less-bitter version of the Negroni cocktail, by substituting Cynar in place of Campari.

Medical Uses

The total antioxidant capacity of artichoke flower heads is one of the highest reported for vegetables. Cynarine is a chemical constituent in *Cynara*. The majority of the cynarine found in artichoke is located in the pulp of the leaves, though dried leaves and stems of artichoke also contain it. It inhibits taste receptors, making water (and other foods and drinks) seem sweet.

Studies have shown artichoke to aid digestion, hepatic and gall bladder function, and raise the ratio of HDL to LDL. This reduces cholesterol levels, which diminishes the risk for arteriosclerosis and coronary heart disease. Aqueous extracts from artichoke leaves have also been shown to reduce cholesterol by inhibiting HMG-CoA reductase and having a hypolipidemic influence, lowering blood cholesterol. Artichoke contains the bioactive agents apigenin and luteolin. *C. scolymus* also seems to have a bifidogenic effect on beneficial gut bacteria. Artichoke leaf extract has proved helpful for patients with functional dyspepsia, and may ameliorate symptoms of irritable bowel syndrome.

Growing Artichokes

Native to the Mediterranean, growing artichokes (*Cynara scolymus*) requires cool nights and warm days. Aside from providing delicious, tender thistles for the table, the plants themselves are gorgeous! They grow to 5 feet across and almost as high with beautiful gray fuzzy foliage.

Growing artichokes as annuals that bear edible buds their first season requires an early start, but properly handled artichoke plants will prosper in a wide range of climates. This guide includes descriptions of the types of globe artichokes and tips for growing them in your backyard garden.

Site Preparation: Each spring, mix compost into your growing area. Artichokes require sandy, fast draining soil and cool temperatures to thrive. They need regular water for an ample harvest, but if you just like the look of the plant and don't want the thistles for your table, they will survive on very little water. Artichokes are susceptible to freezing and do best where the temperature remains constant year round.

How to Plant: Plant artichokes in a location in full sun from bare root stock in January or from container grown stock later in the spring. To grow artichokes in cold winter climates, protect the root with several inches of straw mulch or better yet, grow them in large containers and move to a protected location when the temperature drops. Fertilize with a small amount of all-purpose fish fertilizer. Micronutrients from seaweed extract can be beneficial also.

Harvesting: Artichokes are ready to harvest when the heads are closed tightly and squeak slightly when squeezed. If you wait for them to open, they will be too tough to eat. Search the interior of the plant, the chokes hide in the foliage. Small artichokes can be eaten whole, without removing the inside spiny choke. Artichokes take at least 110-150 days to reach maturity, if planting from seed and 100 days from divisions. Most do not flower until the second year of growth.

Once the harvest is over, cut the plants back to 1-2 inches off the ground to try for a second harvest. New sprouts will form at the base of the plant. At the end of the season, allow the plant to dry out after the leaves begin to turn yellow. Once the foliage has died down and dried, remove it from the plant and put down a layer of organic compost to enrich the soil for next years crop.

Insects and Diseases: Protection from earwigs is mandatory. Also, keep an eye out for aphids, caterpillars, slugs and snails. Use diatomaceous earth or other natural pest control method, if present.

Seed Saving Instructions: Artichoke flower heads are cut when completely open and beginning to show their white seed plumes. Store the flower heads in a dry location away from direct sunlight until dry and brittle. Place one of the dry flower heads in a feed sack or canvas bag on a concrete surface. Pound the base of the blossom with a hammer and allow the down to float out of the bag. The seed is heavy and will remain behind, but should be removed from the bag after each flower is processed to avoid crushing.

Pest Control and Care

Artichoke plants are one of those prehistoric looking specimens that not only create a visual stir in the garden, but also produce delicious globes and unique purple flowers. The plants are relatively easy to grow and become monsters in the landscape. You may on occasion encounter a few problems with artichoke plants while growing them. Artichoke plants under attack may have insect or disease issues. Learn what to look for and how to deal with artichoke pests and proper care of diseased artichokes.

Problems with Artichoke Plants

When you consider the size and roughly serrated, tough leaves of an artichoke plant, it is hard to see what could hurt these mighty thistle relatives. Some of nature's tiniest organisms can have their way with the plant and there are several fungal diseases, which can compromise the plant's health. Young seedlings are prone to damping off, which is a soil borne disease causing it to wilt and die. Flying insect larvae eat all parts of the plant. Sucking insects feed on the sap and creeping snails and slugs make Swiss cheese of the foliage. A watchful gardener can usually nip artichoke plant diseases in the bud. Artichoke plants under attack by insects may need pesticides, but oftentimes the old "pick and squish" method will keep them from much harm.

Artichoke Plant Diseases

The most common diseases affect the leaves and include Then powdery mildew and botrytis. Powdery mildew leaves a white coating on foliage and is caused by fungi, which thrive in moist, warm weather. Botrytis blight is also a fungus but it prefers cool, moist environments and causes the plant to collapse. Curly dwarf virus has a cute name but the effects are detrimental. The virus is transmitted by sucking insects, like aphids, and produces a stunted, sickly plant.

The majority of artichoke plant diseases can be avoided by crop rotation, pest management and avoiding overhead watering. Some diseases, like verticillium wilt, are common on other crop plants like strawberries and lettuces. Avoid planting near these crops to avoid spreading disease. Care of diseased artichokes may involve removal of afflicted plant parts. Healthy, vigorous plants can withstand most artichoke plant diseases.

Artichoke Pests

Some of the most damaging pests are the sucking insects. These include aphids, mites, scab and thrips. They can transmit dangerous artichoke plant diseases as well as reduce the plant's vigor.

Chewing insects reduce the ornamental appeal of the leaves but can also kill the foliage if large numbers attack. Watch for leafhoppers, many forms of moth, cutworms, armyworms and any other larvae. Do not let slow snails and slugs fool you. Their slow crawl up the stout stems of your artichoke can spell disaster for its leaves. Overnight eating patterns will create a lacy tapestry of foliage, which affects the plant's ability to gather solar energy.

When you search for artichoke pests, do not forget to look underneath the huge leaves. The next generation of pest may be in egg form waiting to hatch and get eating. Blast the leaves with water in the morning to remove many insects. Apply horticultural soap or neem oil for heavier infestations and pick off larvae before they can do any serious damage.

2

Asian Greens

Asian greens make an excellent accompaniment to a meal as both a side dish or as part of a stir-fry. And not only are they inexpensive and quick to prepare, they also deliver a nice dose of nutrients.

What are the different varities?

There are a number of difference varieties of Asian greens, with the most common being bok choy, also known as pak choy, Chinese or Nappa cabbage, Chinese broccoli (Gai Lan) and choy sum (Chinese silver beet).

Depending on which part of the plants you use, fast-growing Asian greens can slip into several culinary roles, and all plants are excellent sources of calcium and vitamins A, C and K.

Selecting and storing: When choosing Asian greens, look for crisp stalks with bright green, glossy leaves. Store Asian greens as you would store other greens but for best flavour, use within three days of purchase. In most cases, both the leaves and stalks can be eaten, but be sure to trim the stalks to remove any fibrous outer husk.

Preparing asian vegetables: Asian vegetables should be cooked quickly to seal in their flavour. The most common ways to cook Asian greens are boiling, steaming and stir-frying. Be sure to keep your eye on them while cooking as they can cook in as little as a minute depending on which way you prepare them.

Nutritional features: Most Asian greens come from the Brassica family of vegetables, which are typically rich in vitamin C, beta-carotene, potassium and can provide calcium and magnesium. Asian greens are also a kilojoule bargain, with one cup accounting for a mere 85 kJ as well as providing roughly 2 grams of fibre. In addition to being inexpensive and rich in nutrients, Asian greens are also extremely versatile. They can be used in stir fries, salads, and soups or simply as a side dish.

Bok Choy (*Brassica rapa ssp. chinensis*)

Also known as pak choi or spoon cabbage. In most common types—white-stemmed, green-stemmed, and "soup spoon"—both leaves and stems are edible. Likewise the ever-more-common baby bok choy. One of the most delicately flavoured Asian greens, bok choy is good for stir-frying (in oil, with garlic and soy sauce), braising, or simmering in soups.

Chinese Cabbage (*Brassica rapa ssp. pekinensis*)

The two Chinese cabbages most common here: napa (nothing to do with the California valley), which is barrel-shaped, and michichili, more elongated. Both are mild and crunchy. Use them, thinly sliced and quickly cooked, as a bed for whole steamed fish or barbecued duck; simmer them in soup; or mix with meat as a filling for Chinese dumplings.

Chinese cabbage (*Brassica rapa*, subspecies *pekinensis* and *chinensis*) can refer to two distinct varieties of Chinese leaf vegetables often used in Chinese cuisine: *Pekinensis* (napa cabbage) and *Chinensis* (bok choi).

These vegetables are both variant cultivars or subspecies of the turnip and belong to the same genus as such Western staples as cabbage, broccoli, and cauliflower. Both have many variations in name, spelling, and scientific classification—especially the bok choy (*B. rapa chinensis*) variety.

There is evidence that the turnip was domesticated in India for its oil-bearing seeds before the 15th century BCE. Unlike the European turnip, however, the cultivation of the Chinese cabbage has focused on its leaves instead of its roots.

It was principally grown in the Yangtze River Delta region, but the Ming Dynasty naturalist Li Shizhen popularized it by bringing attention to its medicinal qualities. The variant cultivated in Zhejiang around the 14th Century was brought north and the northern harvest of napa cabbage soon exceeded the southern one. These were then exported back south along the Grand Canal to Hangzhou and traded by sea as far south as Guangdong.

The Napa cabbage became a Manchurian staple for making *suan cai*, the Chinese sauerkraut. In Korea, this developed into kimchi. The vegetable spread to Japan following its wars in Manchuria during the Russo-Japanese War in the early 20th Century. Chinese cabbage is now commonly found in markets throughout the world, catering both to the Chinese diaspora and to northern markets who appreciate its resistance to cold.

Varieties

There are two distinctly different groups of *Brassica rapa* used as leaf vegetables in China, and a wide range of varieties within these two groups. The binomial name *B. campestris* is also used.

Pekinensis

This group is the more common of the two, especially outside Asia; names such as napa cabbage, *Baguio petsay* or *petsay wombok* (Tagalog); *Chinese white cabbage*; "wong a pak" (Hokkien, Fujianese); *baechu* (Korean), *wongbok* and *hakusai* usually refer to members of this group. *Pekinensis* cabbages have broad green leaves with white petioles, tightly wrapped in a cylindrical formation and usually forming a compact head. As the group name indicates, this is particularly popular in northern China around Beijing (Peking).

Chinensis

Chinensis varieties do not form heads; instead, they have smooth, dark green leaf blades forming a cluster reminiscent of mustard or celery. *Chinensis* varieties are popular in southern China and Southeast Asia. Being winter-hardy, they are increasingly grown in Northern Europe. This group was originally classified as its own species under the name *B. chinensis* by Linnaeus.

Chinensis *Spelling and Naming Variations*

Other than the ambiguous term "Chinese cabbage," the most widely used name in North America for the *chinensis* variety is *bok choy* (from Cantonese, literally "white vegetable"; also spelled *Pak choi*, *Bok choi*, and *Pak choy*). In the UK, Australia, South Africa, and other Commonwealth Nations, the term *Pak choi* is used. Less commonly, the descriptive English names *Chinese chard*, *Chinese mustard*, *celery mustard*, and *Spoon cabbage* are also employed.

In Australia, the New South Wales Department of Primary Industries has redefined many of these names to refer to specific cultivars. In addition, they have introduced the word *buk choy* to refer to a specific kind of cabbage distinct from *pak choy*.

In China, three terms are commonly used for this vegetable: the majority of Chinese (about 955 million) speak Mandarin, and for them the term is literally "oil vegetable", since most of the cooking oil in China is extracted from the seed of this plant; Shanghainese speakers (about 90 million in eastern China) use the term R—"blue-green vegetable"; although the term pronounced "baak choi" in Cantonese, the same characters are pronounced "bái cài" by Mandarin speakers and used as the name for Napa cabbage which they call as "Chinese cabbage" when speaking English.

Nutritional Value

Pak choi contains a high amount of vitamin A per 4 oz. of serving - about 3500 IU. Pak choi also contains approximately 50 mg of vitamin C per 4 oz. serving.

Toxic Effects

Pak choi contains glucosinolates. These compounds have been reported to prevent cancer in small doses, but, like many substances, can be toxic to humans in large doses, particularly to people who are already seriously ill. In 2009, an elderly diabetic woman who had been consuming 1 to 1.5 kg of raw Pak choi per day, in an attempt to treat her diabetes, developed hypothyroidism, for reasons relating to her diabetes, resulting in myxedema coma.

Choy Sum (*Brassica rapa var. parachinensis*)

Also known as yu choy; one type is called Chinese oil vegetable. The slightly bitter stems are more tender than the leaves, although all parts of the vegetable, even the tiny yellow flowers, are edible. Thick stems should be trimmed, peeled, then steamed or poached. Stir-fried, choy sum stands up well to the assertive flavours of garlic, rice wine, and chiles.

Chinese Flowering Cabbage (*Brassica rapa* var. *parachinensis* or *Brassica chinensis* var. *parachinensis*) is a member of the *Brassica*

genus and the Brassicaceae (mustard) family. Choy sum is mainly characterized by yellow flowers with each flower having four yellow, oval to round petals with six stamens on fleshy, erect stems which are 0.5 to 1 cm in diametre and 15 to 20 cm tall with light to dark green, and are oval (becomes acuminate shaped, or basal-shaped near the flowering stage) with slightly serrated margins leaves, which never forms compact heads like the cabbage. Fruits can develop out of cross-pollination or self-pollination, and are silique structured, that opens at maturity through dehiscence or drying to bare open to brown or black seeds that are small and round in shape.

A single pod can bear up to 4 to 46 seeds. The height of the plant varies greatly, ranging from 10 to 40 centimetres depending on the growing conditions and the variety. Flowering usually appears when there are about 7 to 8 leaves on the plant or about 20 centimetres tall. The bulk of the root system is found within a depth of 12 cm and is confined to a radius of 12 cm.

The whole plant is overall an annual, herbaceous plant, rarely perennial, rarely growing into subshrubs. The whole plant consists of a simple or branched (when it is near the flowering stage), leafy structure. It grows best in soil with a pH range of minimum pH level at 5.6, maximum pH level at 7.5.

Chrysanthemum Greens (*Chrysanthemum coronarium*)

Often called garland chrysanthemums or crown daisies. Botanically different from other chrysanthemums. Leaves are flat or serrated.

Tangy flavour. Becomes bitter if overcooked. Young leaves are quickly blanched to darken them, then coated with oil and added to salads. The Japanese use the leaves in tempura. The Chinese garnish soup with the petals.

Chrysanthemum coronarium is a species of flowering plant in the aster family, Asteraceae. It is native to the Mediterranean and East Asia. It is used as a leaf vegetable. English language common names include garland chrysanthemum, chrysanthemum greens, edible chrysanthemum, chop suey green, crown daisy, and Japanese-green.

Characteristics

A leafy herb, the garland chrysanthemum is one of the few annual plants in its genus. It has yellow ray florets grouped in small flower heads and aromatic, bipinnately lobed leaves. The vegetable grows very well in mild or slightly cold climates, but will go quickly into premature flowering in warm summer conditions. Seeds are sown in early spring and fall.

"The plant is rich in minerals and vitamins with potassium concentrations at 610 mg/100 g and carotene at 3.4 g/100 g in edible portions. In addition, the plant contains various antioxidants (in stem, leaf, and root tissues) that have potential long-term benefits for human health, although toxic (dioxin) properties have also been observed. Extracts from *C. coronarium* var. *spatiosum* have been shown to inhibit growth of *Lactobacillus casei*, a beneficial human intestinal bacterium."

Culinary Uses

The plant's greens are used in many Asian cuisines. They appear in Cantonese dishes and Hong Kong cuisine in stews, casseroles, and hotpots. The leaves are also an important ingredient in Taiwanese oyster omelettes and, when young, are used along with stems to flavour soup and stir-fry. In Japan, it is used in *nabemono*. Korean cookery uses the greens in soups, stews, and alone as a side dish or (*banchan*). In a hotpot, it is added at the last moment to the pot to avoid overcooking.

In Crete, a variety of the species called *mantilida* has its tender shoots eaten raw or steamed by the locals

Mizuna (*Brassica rapa ssp. nipposinica var. laciniata*)

Both the Japanese name, mizuna, and the Mandarin Chinese name, shui cai, mean "water vegetable" (the stalks are very juicy). Used frequently in Japanese cooking. The dark green, feathery leaves are peppery. Baby leaves are often used in field green mixtures; mature ones are excellent tossed with sautéed chicken or roasted root vegetables.

Brassica rapa L. is a plant consisting of various widely cultivated subspecies including the turnip (a root vegetable); the mizuna, napa cabbage, and cime di rapa (leaf vegetables); and turnip rape (*Brassica rapa* subsp. *oleifera*, an oilseed which has many common names, including [*annual*] *turnip rape*, *canola*, *field mustard*, *bird rape*, *keblock*, and *colza*). The oilseed is sometimes confused with rapeseed oil, which however comes from a different *Brassica* species.

In the 18th century the turnip and the oilseed-producing variants were seen as being different species by Carolus Linnaeus who named them *B. rapa* and *B. campestris*. 20th-century taxonomists found that the plants were cross fertile and thus belonged to the same species. Since the turnip had been named first by Linnaeus, the name *Brassica rapa* was adopted.

The oilseeds known as canola are sometimes particular varieties of *Brassica rapa* (termed Polish Canola) but mostly related species *Brassica napus* and *Brassica juncea*.

Many butterflies, including small white pollinate the *B. rapa* flowers.

Green Mustard (*Brassica juncea*)

Also called Chinese mustard greens and leaf mustard, this is among the most pungent of Asian greens. There are numerous sub-varieties, with leaves taking various shapes. Steamed for five minutes or so and dressed with a little oil, it makes a perfect side dish. Its flavour and texture work well with oyster sauce and with the richness of pork. It is also good pickled.

Brassica juncea, mustard greens, Indian mustard, Chinese mustard, or leaf mustard is a species of mustard plant. Subvarieties include southern giant curled mustard, which resembles a headless cabbage such as kale, but with a distinct horseradish-mustard flavour. It is also known as green mustard cabbage.

Uses

Food: The leaves, the seeds, and the stem of this mustard variety are edible. The plant appears in some form in African, Italian, Indian, Chinese, Japanese, Korean, and soul food cuisine. Cultivars of *B. juncea* are grown as greens, and for the production of oilseed.

In Russia, this is the main variety grown for production of mustard oil, which after refining is considered one of the best vegetable oils. It is widely used in canning, baking and margarine production in Russia, and the majority of table mustard there is also made from this species of mustard plant.

The leaves are used in African cooking, and leaves, seeds, and stems are used in Indian cuisine, particularly in mountain regions of Nepal, as well as in the Punjab cuisine of India and Pakistan, where a famous dish called *sarson da saag* (mustard greens) is prepared. *B. juncea* subsp. *tatsai*, which has a particularly thick stem, is used to make the Indian pickle called *achar*, and the Chinese pickle *zha cai*. The mustard made from the seeds of the *B. juncea* is called brown mustard. The leaves (*raai* in Gujarati) are used in many Indian dishes.

The Gorkhas of Darjeeling and Sikkim prepare pork with mustard greens (also called *rayo* in Nepali). It is usually eaten with relish with steamed rice, but could also be eaten with *chapati* (griddle breads).

Brassica juncea is more pungent than the closely related *Brassica oleracea* greens (kale, cabbage, collard greens, et cetera), and is frequently mixed with these milder greens in a dish of "mixed greens", which may include wild greens such as dandelion. As with other greens in soul food cooking, mustard greens are generally flavoured by being cooked for a long period with ham hocks or other smoked pork products. Mustard greens are high in vitamin A and vitamin K.

Chinese and Japanese cuisines also make use of mustard greens. In Japanese cuisine it is known as Takana and is often pickled and used as filling in onigiri or as a condiment. A large variety of *B. juncea* cultivars are used, including *zha cai*, *mizuna*, *takana* (var. *integlofolia*), *juk gai choy*, and *xuelihong*.

Asian mustard greens are most often stir-fried or pickled. A Southeast Asian dish called *asam gai choy* or *kiam chai boey* is often made with leftovers from a large meal. It involves stewing mustard greens with tamarind, dried chillies and leftover meat on the bone.

Figure: *Cantonese-style braised mustard greens, with wolfberries*

Food Supplement

B. juncea can hyperaccumulate cadmium and many other soil trace elements. Specially cultured, it can be used as a selenium, chromium, iron and zinc food supplement.

Green Manure

Figure: *A bunch of fresh mustard greens from the United States*

Vegetable growers sometimes grow mustard as a green manure. Its main purpose is to act as a mulch, covering the soil to suppress weeds between crops. If grown as a green manure, the mustard plants are cut down at the base when sufficiently grown, and left to wither on the surface, continuing to act as a mulch until the next crop is due for sowing, when the mustard is dug in. In the UK, summer and autumn-sown mustard is cut down from October. April sowings can be cut down in June, keeping the ground clear for summer-sown crops. One of the disadvantages of mustard as a green manure is its propensity to harbor club root.

Phytoremediation

This plant is used in phytoremediation to remove heavy metals, such as lead, from the soil in hazardous waste sites because it has a higher tolerance for these substances and stores the heavy metals in its cells. The plant is then harvested and disposed of properly. This method is easier and less expensive than traditional methods for the removal of heavy metals. It also prevents erosion of soil from these sites preventing further contamination.

Red Mustard (*Brassica juncea*)

More purple than red in colour, this is one of the newer mustard varieties. Its powerful flavour, reminiscent of wasabi, fades if it's overcooked. Used more in Japanese and Korean cooking than in the Chinese kitchen. In the U.S., leaves can be served raw as a decorative, edible accompaniment to sushi, or added to soups just before serving (to preserve the leaf's character).

Red mustard greens AKA Osaka Purple mustard is a member of the Brassica family, collectively known as cruciferous vegetables, along with turnips, broccoli and cabbage. Red mustard greens are grown for fresh eating, as a cover crop with pesticide qualities, for feedstock and as dual purpose ornamental greens in edible gardens.

Description/Taste

Red mustard greens have broad savoyed leaves with a violet purple overlay and purple-green variegation. The leaves grow to 12 inches in length and up to six inches in width. Red mustard's leaves are succulent and tender, the stems, crisp and crunchy. Red mustard greens elicit a savory,spicy and nuanced peppery flavour.

Applications

Red mustard is a common salad green, pot herb, braising and pickling green in a diversity of cuisines from Asia to Europe to South

America. Red mustard pairs well with poultry, legumes, sausages, pork, grilled fish, garlic, creamy sauces and fresh cheeses, nutty oils, light vinegars, citruses such as grapefruit and lemon, mushrooms, fennel, cucumbers, tomatoes, avocados, cumin, basil, mint, pomegranates and stone fruits such as peaches and cherries.

Geography/History

Red mustard greens are Chinese in origin, eventually become a naturalized common green in Japan. Taxonomists identify as many as seventeen subgroups of mustard greens that can differ sharply in heat, flavour, and appearance. Differentiation in soil types and temperatures can affect the flavour and heat level of mustard greens. All Red mustard varieties prefer cool climates for growing with full sun and rich soil with temperatures below 68°F. Frost is tolerable, but freezing temperatures will kill crops.

Pea Shoots (*Pisum stivum*)

From the *Leguminosae* family. Correctly referred to as pea pod leaves, these are the tendrils and top few leaves of the snow pea plant. An important ingredient in Shanghai and Vietnamese cooking. Pinch leaves from stems, and use the tenderest tips. The delicate pea flavour comes through when they're used raw in salads, steamed, or stir-fried for a minute with a little ginger.

The pea is most commonly the small spherical seed or the seed-pod of the pod fruit *Pisum sativum*. Each pod contains several peas. Peapods are botanically a fruit, since they contain seeds developed from the ovary of a (pea) flower. The name is also used to describe other edible seeds from the Fabaceae such as the pigeon pea (*Cajanus cajan*), the cowpea (*Vigna unguiculata*), and the seeds from several species of *Lathyrus*.

P. sativum is an annual plant, with a life cycle of one year. It is a cool season crop grown in many parts of the world; planting can take place from winter to early summer depending on location. The average pea weighs between 0.1 and 0.36 grams.

The immature peas (and in snow peas the tender pod as well) are used as a vegetable, fresh, frozen or canned; varieties of the species typically called field peas are grown to produce dry peas like the split pea shelled from the matured pod. These are the basis of pease porridge and pea soup, staples of medieval cuisine; in Europe, consuming fresh immature green peas was an innovation of Early Modern cuisine.

The wild pea is restricted to the Mediterranean basin and the Near East. The earliest archaeological finds of peas date from the late neolithic era of current Greece, Syria, Turkey and Jordan. In Egypt, early finds date from *ca.* 4800–4400 BC in the Nile delta area, and from *ca.* 3800–3600 BC in Upper Egypt. The pea was also present in Georgia in the 5th millennium BC. Farther east, the finds are younger. Peas were present in Afghanistan *ca.* 2000 BC, in Harappa, Pakistan, and in northwest India in 2250–1750 BC. In the second half of the 2nd millennium BC, this pulse crop appears in the Gangetic basin and southern India.

Culture

Plant as a cool season crop, ideally between 55 degrees and 65 degrees F. Sow seeds closely where they are to grow in fertile, moisture-retentive, neutral soil. Podded varieties should be staked or grown on a trellis but pea-shoot varieties can be allowed to creep along the ground.

Noteworthy Characteristics

Peas are hardy annuals native to Eurasia that have been grown since 7,000 B.C. The common garden pea, *Pisum sativum* var. *sativum*, has a fibrous pod that is not edible. The seeds are harvested when almost mature and eaten fresh as shelled peas or the pods can be left to mature when the mature seeds are harvested and used dried. *Pisum sativum* var. *macrocarpon* lacks the fibres in the inner lining found in the common pea. Hence, the pod is edible as well as the tender peas inside. This group includes the edible-podded, sugar, snow and snap peas. The young tips, called pea shoots, of any of the varieties of *Pisum sativum* may be harvested and cooked as a pot herb. Pea shoots are a delicacy in Chinese cuisine. The smooth-seeded types are more cold hardy than the wrinkled-seeded varieties.

Garden Uses

Depending upon the cultivar chosen, peas can be grown for their sweet shelled peas, as edible-podded, as snap peas, as Asian snow peas, or for their edible shoots. Snap and snow peas work well in Asian stir-frys and the edible shoots are excellent raw or cooked in soup, steamed, or stir fried. Eat raw with lemon juice dripped over them, wilted with olive oil and garlic, steamed or sautéed in butter with salt and butter. Substitute for spinach in Florentine seafood dishes. Bake in white sauce au gratin, or serve with a cream cheese or mustard sauce.

Pests and Diseases

A variety of diseases affect peas through a number of pathogens, including insects, viruses, bacteria and fungi. In particular, virus disease of peas has worldwide economic importance.

Powdery mildew and slugs may be occasional problems. Seeds can also rot if planted too early in damp, cold ground.

Additionally, insects such as the pea leaf weevil (*Sitona lineatus*) can damage peas and other pod fruits. The pea leaf weevil is native to Europe, but has spread to other places such as Alberta, Canada. They are about 3.5 millimetres (0.14 in)—5.5 millimetres (0.22 in) long and are distinguishable by three light-coloured stripes running length-wise down the thorax. The weevil larvae feed on the root nodules of pea plants, which are essential to the plants' supply of nitrogen, and thus diminish leaf and stem growth. Adult weevils feed on the leaves and create a notched, "c-shaped" appearance on the outside of the leaves.

Tatsoi (*Brassica rapa var. rosularius or atrovirens*)

A ground-hugging member of the bok choy family, also known as rosette bok choy for its conformation: The round, thick, very dark green leaves grow in tight, concentric circles like rose petals. With its slightly bitter flavour, tatsoi is excellent raw (when young) in salads, and adds spark to Asian-style soups when tossed in at the last minute.

Tatsoi (*Brassica narinosa* or *Brassica rapa var. rosularis*), also called Spinach mustard, Spoon mustard, or Rosette bok choy, is an Asian variety of *Brassica rapa* grown for greens. This plant has become popular in North American cuisine as well, and is now grown throughout the world.

The plant has dark green spoon-shaped leaves which form a thick rosette. It has a soft creamy texture and has a subtle yet distinctive flavour.

It can be grown to harvestable size in 45–50 days, and can withstand temperatures down to –10°C (15°F). Tatsoi can be harvested even from under the snow.

Asparagus

Plant once, harvest for years: A well-maintained bed of this sweet, slender veggie will stay productive for up to 15 years, and, with its vibrant, ferny foliage, asparagus makes an excellent ornamental.

Asparagus officinalis is a spring vegetable, a flowering perennial plant species in the genus *Asparagus*. It was once classified in the lily family, like its *Allium* cousins, onions and garlic, but the Liliaceae have been split and the onion-like plants are now in the family Amaryllidaceae and asparagus in the Asparagaceae. *Asparagus officinalis* is native to most of Europe, northern Africa and western Asia, and is widely cultivated as a vegetable crop.

Biology

Asparagus is a herbaceous, perennial plant growing to 100–150 centimetres (39–59 in) tall, with stout stems with much-branched feathery foliage. The "leaves" are in fact needle-like cladodes (modified stems) in the axils of scale leaves; they are 6–32 mm (0.24–1.26 in) long and 1 mm (0.039 in) broad, and clustered 4–15 together. The root system is adventitious and the root type is fasciculated. The flowers are bell-shaped, greenish-white to yellowish, 4.5–6.5 mm (0.18–0.26 in) long, with six tepals partially fused together at the base; they are produced singly or in clusters of two or three in the junctions of the branchlets. It is usually dioecious, with male and female flowers on separate plants, but sometimes hermaphrodite flowers are found. The fruit is a small red berry 6–10 mm diametre, which is poisonous to humans.

Plants native to the western coasts of Europe (from northern Spain north to Ireland, Great Britain, and northwest Germany) are treated as *Asparagus officinalis* subsp. *prostratus* (Dumort.) Corb., distinguished by its low-growing, often prostrate stems growing to only 30–70 cm (12–28 in) high, and shorter cladodes 2–18 mm (0.079–0.709 in) long. It is treated as a distinct species, *Asparagus prostratus* Dumort, by some authors. A remarkable adaptation is the edible asparagus, while in the Macaronesian Islands several species, (*A. umbellatus*, *A. scoparius*, etc.), are preserved the original form, a leafy vine; in the Mediterranean, the asparagus genus has evolved into thorny species.

History

Asparagus has been used as a vegetable and medicine, owing to its delicate flavour, diuretic properties, and more. It is pictured as an offering on an Egyptian frieze dating to 3000 BC. In ancient times, it was also known in Syria and in Spain. Greeks and Romans ate it fresh when in season, and dried the vegetable for use in winter; Romans even froze it high in the Alps, for the Feast of Epicurus. Emperor Augustus created the "Asparagus Fleet" for hauling the vegetable, and coined the expression "faster than cooking asparagus" for quick action. A recipe for cooking asparagus is in the oldest surviving book of recipes, Apicius's third-century AD *De re coquinaria*, Book III.

The ancient Greek physician Galen (prominent among the Romans) mentioned asparagus as a beneficial herb during the second century AD, but after the Roman empire ended, asparagus drew little medieval attention. until al-Nafzawi's *The Perfumed Garden*. That piece of writing celebrates its (scientifically unconfirmed) aphrodisiacal power, a supposed virtue that the Indian *Ananga Ranga* attributes to "special phosphorus elements" that also counteract fatigue. By 1469, asparagus was cultivated in French monasteries. Asparagus appears to have been hardly noticed in England until 1538, and in Germany until 1542.

The finest texture and the strongest and yet most delicate taste is in the tips. The *points d'amour* ("love tips") were served as a delicacy to Madame de Pompadour. Asparagus became available to the New World around 1850, in the United States.

Uses

Only young asparagus shoots are commonly eaten: once the buds start to open ("ferning out"), the shoots quickly turn woody.

Water makes up 93% of Asparagus's composition. Asparagus is low in calories and is very low in sodium. It is a good source of vitamin B_6, calcium, magnesium and zinc, and a very good source of dietary fibre, protein, beta-carotene, vitamin C, vitamin E, vitamin K, thiamin, riboflavin, rutin, niacin, folic acid, iron, phosphorus, potassium, copper, manganese and selenium, as well as chromium, a trace mineral that enhances the ability of insulin to transport glucose from the bloodstream into cells. The amino acid asparagine gets its name from asparagus, as the asparagus plant is relatively rich in this compound.

The shoots are prepared and served in a number of ways around the world, typically as an appetizer or vegetable side dish. In Asian-style cooking, asparagus is often stir-fried. Cantonese restaurants in the United States often serve asparagus stir-fried with chicken, shrimp, or beef. Asparagus may also be quickly grilled over charcoal or hardwood embers. It is also used as an ingredient in some stews and soups. In recent years asparagus eaten raw, as a component of a salad, has regained popularity.

Asparagus can also be pickled and stored for several years. Some brands label shoots prepared this way as "marinated".

Stem thickness indicates the age of the plant, with the thicker stems coming from older plants. Older, thicker stalks can be woody, although peeling the skin at the base removes the tough layer. Peeled asparagus will however poach much faster. The bottom portion of asparagus often contains sand and dirt, so thorough cleaning is generally advised before cooking.

Green asparagus is eaten worldwide, though the availability of imports throughout the year has made it less of a delicacy than it once was. In Europe, however, the "asparagus season is a highlight of the foodie calendar"; in the UK this traditionally begins on 23 April and ends on Midsummer Day. As in continental Europe, due to the short growing season and demand for local produce, asparagus commands a premium price.

White Asparagus in Continental Northwestern Europe

Asparagus is very popular in the Netherlands, Spain, France, Poland, Belgium, Germany, Italy and Switzerland, and is almost exclusively white—else, it is specified by the local language term for "green asparagus". White asparagus is the result of applying a blanching technique while the asparagus shoots are growing. Compared to green asparagus, the locally cultivated so-called "white gold" or "edible ivory" asparagus, also referred to as "the royal vegetable", is less bitter and much more tender. Freshness is very important, and the lower ends of white asparagus must be peeled before cooking or raw consumption.

To cultivate white asparagus, the shoots are covered with soil as they grow, i.e. hilling; without exposure to sunlight no photosynthesis starts, so the shoots remain white in colour. Only seasonally on the menu, asparagus dishes are advertised outside many restaurants, usually from late April to June. For the French style, asparagus is often boiled or steamed and served with hollandaise sauce, melted butter or olive oil, Parmesan cheese or mayonnaise. Tall, narrow asparagus cooking pots allow the shoots to be steamed gently, their tips staying out of the water.

During the German *Spargelsaison* or *Spargelzeit* ("asparagus season" or "asparagus time"), the asparagus season that traditionally finishes on 24 June, roadside stands and open-air markets sell about half of the country's white asparagus consumption.

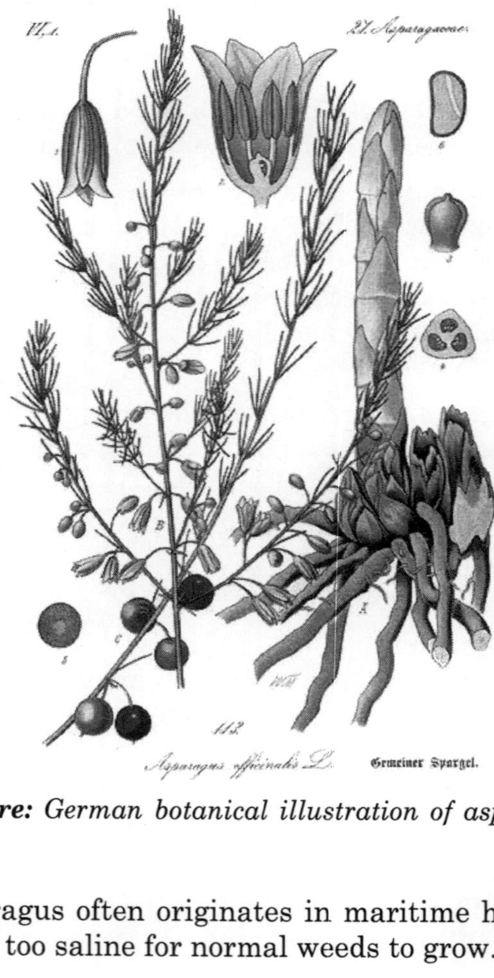

Figure: *German botanical illustration of asparagus*

Cultivation

Since asparagus often originates in maritime habitats, it thrives in soils that are too saline for normal weeds to grow. Thus, a little salt was traditionally used to suppress weeds in beds intended for asparagus; this has the disadvantage that the soil cannot be used for anything else. Some places are better for growing asparagus than others. The fertility of the soil is a large factor. "Crowns" are planted in winter, and the first shoots appear in spring; the first pickings or "thinnings" are known as sprue asparagus. Sprue has thin stems.

A new breed of "Early Season Asparagus" that can be harvested two months earlier than usual was announced by a UK grower in early 2011. This variety does not need to lie dormant and blooms at 7 °C (45 °F) rather than the usual 9 °C (48 °F).

The blanching of white asparagus is obtained by the labour intensive hilling cultivation method, to distinguish its gastronomical qualities from those of the green plant, which is the same botanical variety.

Purple asparagus differs from its green and white counterparts, having high sugar and low fibre levels. Purple asparagus was originally developed in Italy and commercialized under the variety name "Violetto d' Albenga". Since then, breeding work has continued in the United States and New Zealand.

Companion Planting

Asparagus is a useful companion plant for tomatoes. The tomato plant repels the asparagus beetle, as do several other common companion plants of tomatoes. Meanwhile, asparagus may repel some harmful root nematodes that affect tomato plants.

Commercial Production

The top asparagus importers (2004) were the United States (92,405 tonnes), followed by the European Union (external trade) (18,565 tonnes), and Japan (17,148 tonnes).

China is the world's largest producer: in 2010 (6,960,357 tonnes), at a large distance followed by Peru (335,209 tonnes), and Germany (92,404 tonnes). U.S. production was concentrated in California, Michigan and Washington. The annual production for white asparagus in Germany is 57,000 tonnes (61% of consumer demand).

Celebrations

The green crop is significant enough in California's Sacramento-San Joaquin River Delta region that the city of Stockton holds a festival every year to celebrate it, as does the city of Hart, Michigan, complete with a parade and asparagus queen. The Vale of Evesham in Worcestershire is heralded as the largest producer within Northern Europe, celebrating like Stockton, with a week-long festival every year involving auctions of the best crop and locals dressing up as spears of asparagus as part of the British Asparagus Festival.

Many German cities hold an annual *Spargelfest* (asparagus festival) celebrating the harvest of white asparagus. Schwetzingen claims to be the "Asparagus Capital of the World" and during its festival an Asparagus Queen is crowned. The Bavarian city of Nuremberg feasts a week long in April, with a competition to find the fastest asparagus peeler in the region. This usually involves generous amounts of the local wines and beer being consumed to aid the spectators' appreciative support.

Vernacular Names and Etymology

Asparagus officinalis is widely known simply as "asparagus", and may be confused with unrelated plant species also known as "asparagus", such as *Ornithogalum pyrenaicum* known as "Prussian asparagus" for its edible shoots.

The English word "asparagus" derives from classical Latin, but the plant was once known in English as *sperage*, from the Medieval Latin *sparagus*. This term itself derives from the Greek *aspharagos* or *asparagos*, and the Greek term originates from the Persian *asparag*, meaning "sprout" or "shoot".

Asparagus was also corrupted in some places to "sparrow grass"; indeed, the Oxford English Dictionary quotes John Walker as having written in 1791 that "*Sparrow-grass* is so general that *asparagus* has an air of stiffness and pedantry". In Gloucestershire and Worcestershire, it is also known simply as "grass", and young plants too small to cut are called "pru".

Another known colloquial variation of the term, most common in parts of Texas, is "aspar grass" or "asper grass". In the Midwest United States and Appalachia, "spar grass" is a common colloquialism. Asparagus is commonly known in fruit retail circles as "Sparrows Guts", etymologically distinct from the old term "sparrow grass".

The Sanskrit name of Asparagus is *shatavari* and it has been historically used in India as a part of Ayurvedic medicines. In Kannada, it is known as *ashadhi*, *majjigegadde* or *sipariberuballi*.

In China it is known as lu sun, in Thailand as *no mai farang*, and in Vietnam as *măng tây* which literally mean "European bamboo shoots" and "Western bamboo shoots", respectively. The green asparagus is commonly used in Chinese-American cuisine and Thai cuisine.

In Turkish, asparagus is known as "ku°konmaz," literally "bird can't land," in reference to the shape of the plant.

Effects on Urine

The effect of eating asparagus on the urine has long been observed:

> *"[Asparagus] cause a powerful and disagreeable smell in the urine, as every Body knows."* (Treatise of All Sorts of Foods, *Louis Lemery, 1702)*
>
> *"asparagus... affects the urine with a foetid smell (especially if cut when they are white) and therefore have been suspected by some physicians as not friendly to the kidneys; when they are older, and begin to ramify, they lose this quality; but then they are not so agreeable." ("An Essay Concerning the Nature of Aliments," John Arbuthnot, 1735)*
>
> *"A few Stems of Asparagus eaten, shall give our Urine a disagreable Odour..." ("Letter to the Royal Academy of Brussels," Benjamin Franklin, c. 1781)*
>
> *Asparagus "...transforms my chamber-pot into a flask of perfume." Marcel Proust (1871–1922)*

There is debate about whether all—or only some—people produce the smell, and whether all (or only some) people identify the smell. It was originally thought this was because some of the population digested asparagus differently from others, so some people excreted odorous urine after eating asparagus, and others did not. In the 1980s three studies from France, China and Israel published results showing that producing odorous urine from asparagus was a common human characteristic. The Israeli study found that from their 307 subjects all of those who could smell 'asparagus urine' could detect it in the urine of anyone who had eaten asparagus, even if the person who produced it could not detect it. However, a 2010 study found variations in both production of odorous urine and the ability to detect the odour, but that these were not tightly related. It is believed most people produce the odorous compounds after eating asparagus, but only about 22% of the population have the autosomal genes required to smell them.

In 2010, the company 23andMe published a genome-wide association study on whether participants have "ever noticed a peculiar odour when you pee after eating asparagus?" This study pinpointed a single-nucleotide polymorphism (SNP) in a cluster of olfactory genes associated with the ability to detect the odour. While this SNP did not explain all of the difference in detection between people, it provides support for the theory that there are genetic differences in olfactory receptors that lead people to be unable to smell these odorous compounds.

Chemistry

Certain compounds in asparagus are metabolized to yield ammonia and various sulfur-containing degradation products, including various thiols and thioesters, which give urine a characteristic smell.

Some of the volatile organic compounds responsible for the smell are:

- methanethiol
- dimethyl sulfide
- dimethyl disulfide
- bis(methylthio)methane
- dimethyl sulfoxide
- dimethyl sulfone

Subjectively, the first two are the most pungent, while the last two (sulfur-oxidized) give a sweet aroma. A mixture of these compounds form a "reconstituted asparagus urine" odour. This was first investigated in 1891 by Marceli Nencki, who attributed the smell to methanethiol. These compounds originate in the asparagus as asparagusic acid and its derivatives, as these are the only sulfur-containing compounds unique to asparagus. As these are more present in young asparagus, this accords with the observation that the smell is more pronounced after eating young asparagus. The biological mechanism for the production of these compounds is less clear.

The onset of the asparagus urine smell is remarkably rapid. The smell has been reported to be detectable 15 to 30 minutes after ingestion.

3

Broccoli and other Flower Vegetables

Tasty in each of its many varieties, broccoli is easier to grow than its relatives cauliflower and Brussels sprouts, and can produce bountiful crops for even novice gardeners.

Broccoli is an edible green plant in the cabbage family, whose large flowering head is used as a vegetable.

The word broccoli comes from the Italian plural of *broccolo*, which means "the flowering crest of a cabbage", and is the diminutive form of *brocco*, meaning "small nail" or "sprout". Broccoli is often boiled or steamed but may be eaten raw.

Broccoli is classified in the Italica cultivar group of the species *Brassica oleracea*. Broccoli has large flower heads, usually green in colour, arranged in a tree-like structure on branches sprouting from a thick, edible stalk. The mass of flower heads is surrounded by leaves. Broccoli resembles cauliflower, which is a different cultivar group of the same species.

Broccoli is a result of careful breeding of cultivated leafy cole crops in the Northern Mediterranean in about the 6th century BC. Since the Roman Empire, broccoli has been considered a uniquely valuable food among Italians.

Broccoli was brought to England from Antwerp in the mid-18th century by Peter Scheemakers. Broccoli was first introduced to the United States by Italian immigrants but did not become widely known there until the 1920s.

Nutrition

Broccoli is high in vitamin C and dietary fibre. It also contains multiple nutrients with potent anti-cancer properties, such as diindolylmethane and small amounts of selenium. A single serving provides more than 30 mg of vitamin C and a half-cup provides 52 mg of vitamin C. The 3,3'-Diindolylmethane found in broccoli is a potent modulator of the innate immune response system with anti-viral, anti-bacterial and anti-cancer activity. Broccoli also contains the compound glucoraphanin, which can be processed into an anti-cancer compound sulforaphane, though the anti-cancer benefits of broccoli are greatly reduced if the vegetable is boiled. Broccoli is also an excellent source of indole-3-carbinol, a chemical which boosts DNA repair in cells and appears to block the growth of cancer cells.

Boiling broccoli reduces the levels of suspected anti-carcinogenic compounds, such as sulforaphane, with losses of 20–30% after five minutes, 40–50% after ten minutes, and 77% after thirty minutes. However, other preparation methods such as steaming, microwaving, and stir frying had no significant effect on the compounds.

Broccoli has the highest levels of carotenoids in the brassica family. It is particularly rich in lutein and also provides a modest amount of beta-carotene.

Consumption of broccoli has been linked to reduced testicular volume in adult males, though more studies are being undertaken to assess the statistical significance of these results.

Varieties

Figure: *Broccoli plants in a nursery*

There are three commonly grown types of broccoli. The most familiar is Calabrese broccoli, often referred to simply as "broccoli", named after Calabria in Italy. It has large (10 to 20 cm) green heads and thick stalks. It is a cool season annual crop. Sprouting broccoli has a larger number of heads with many thin stalks. Purple cauliflower is a type of broccoli sold in southern Italy, Spain, and the United Kingdom. It has a head shaped like cauliflower, but consisting of tiny flower buds. It sometimes, but not always, has a purple cast to the tips of the flower buds.

Other cultivar groups of *Brassica oleracea* include cabbage (Capitata Group), cauliflower and Romanesco broccoli (Botrytis Group), kale and collard greens (Acephala Group), kohlrabi (Gongylodes Group), Brussels sprouts (Gemmifera Group), and Chinese broccoli (Alboglabra Group). Rapini, sometimes called "broccoli raab" among other names, forms similar but smaller heads, and is actually a type of turnip (*Brassica rapa*). Broccolini or "Tenderstem broccoli" is a cross between broccoli and Chinese broccoli. Beneforté is a variety of broccoli containing 2-3 times more glucoraphanin that was produced by crossing broccoli with a wild Brassica variety, *Brassica oleracea* var *villosa*.

Production

Cultivation: Broccoli is a cool-weather crop that does poorly in hot summer weather. Broccoli grows best when exposed to an average daily temperature between 18 and 23 °C (64 and 73 °F). When the cluster of flowers, also referred to as a "head" of broccoli, appear in the centre of the plant, the cluster is green. Garden pruners or shears are used to cut the head about an inch from the tip. Broccoli should be harvested before the flowers on the head bloom bright yellow.

While the heading broccoli variety performs poorly in hot weather, mainly due to insect infestation, the sprouting variety is more resistant, though attention must be paid to sucking insects (such as aphids), caterpillars and whiteflies. Spraying of *bacillus thuringiensis* can control caterpillar attacks, while a citronella vase may ward off whiteflies.

Pests

Mostly introduced by accident, "cabbage worms", the larvae of *Pieris rapae*, the small white butterfly are a common pest in broccoli.

Pieris Rapae

Small White: The Small White (*Pieris rapae*) is a small- to medium-sized butterfly species of the Yellows-and-Whites family Pieridae. It is also known as the Small Cabbage White and in New Zealand, simply as White Butterfly. The names "Cabbage Butterfly" and "Cabbage White" can also refer to the Large White. The butterfly can be distinguished by the white colour with small black dots on its wings. They are distinguished from the smaller size and lack of the black band at the tip of their forewings.

It is widespread and populations are found across Europe, North Africa, Asia, and Great Britain. It has also been accidentally introduced to North America, Australia and New Zealand. The caterpillar of this species is seen as a pest for commercial agriculture. Often referred to as the "imported cabbageworm" they are a serious pest to cabbage and other mustard family crops.

The species has a natural range across Europe, Asia, and North Africa. It was accidentally introduced to Quebec, Canada around 1860 and spread rapidly throughout North America.

The species has spread to all of North American life zones from Lower Austral/Lower Sonoran to Canada. Estimates show that a single female of this species might be the progenitor in a few generations of millions. It is absent or scarce in desert and semidesert regions (except for irrigated areas). It is not found north of Canadian life zone, nor on Channel Islands off the coast of southern California.

The Small White will readily lay eggs on both cultivated and wild members of the cabbage family, such as Charlock (*Sinapis arvensis*) and Hedge mustard (*Sisybrium officinale*). *P. rapae* is known to lay eggs singly on the host plant. The egg is characterized by a yellowish colour and 12 longitudinal ridges.

The egg production peaks about a week after adulthood in lab and the female can live up to 3 weeks. Females tend to lay fewer eggs on plants in clumps than on isolated plants. It has been suggested that isothiocyanate compounds in the family Brassicaceae may have been evolved to reduce herbivory by caterpillars of the Small White.

However, this suggestion is not generally accepted because the small white has later been shown to be immune to the isothiocyanate forming reaction due to a specific biochemical adaptation. In contrast, the small white and relatives seem to have evolved as a consequence of this biochemical adaptation to the isothiocyanate-forming glucosinolates.

Traditionally known in the United States as the Imported Cabbage Worm, now more commonly the Cabbage White, the caterpillars are bluish-green, with tiny black pints, a black ring around the spiracles, and a lateral row of yellow dashes, and a yellow middorsal line.

Caterpillars rest on the undersides of the leaves, making them less visible to predators. Although the larval instars have not been fully studied, different instars are easily differentiated simply by comparing sizes, especially the head alone.

During the first and second instar the head is entirely black; third instar has the clypeus yellow but the rest of the head black. In the fourth and fifth instar, there is a dark greenish-yellow dot behind each eye but with rest of the head black. However, the colour of the caterpillar head does not necessarily indicate specific instar, as the time of colour change is not fixed.

In the larval stage, the Small White can be a pest on cultivated cabbages, kale, radish, broccoli, and horseradish. The larva is considered a serious pest for commercial growth of cabbage and other Brassicaceae.

The pupa of *P. rapae* is very close to that of *P. napi*. It is brown to mottled-gray or yellowish, matching the background colour. It has a large head cone, with a vertical abdomen and flared subdorsal ridge. The two (pupa of *P. rapae* and *P. napi*) can be easily distinguished by comparing the proboscis sheath. In *P. rapae*, the proboscis sheath extends far beyond the antennal sheath while in *P. napi*, only a very short distance.

Like its close relative the Large White, the Small White is a strong flyer and the British population is increased by continental immigrants in most years. Adults are diurnal and fly throughout the day, except for early morning and evening. Although there is occasional activity during the later part of the night, it ceases as dawn breaks. Adult *P. rapae* can move many kilometres in individual flights. Adults have been observed to fly as much as 12 km in one flight. On average, a female flies about 0.7 km per day and moves 0.45 km from where she starts. Males patrol all day around host plants to mate with females.

Brussels Sprouts

One of the last veggies to be harvested in early winter, Brussels sprouts bring the gardening year to a delicious close. Growing Brussels sprouts is easy if you plant at the right time and work with vigorous varieties.

The Brussels sprout is a cultivar in the Gemmifera group of cabbages (*Brassica oleracea*), grown for its edible buds. The leafy green vegetables are typically 2.5–4 cm (0.98–1.6 in) in diametre and look like miniature cabbages. The Brussels sprout has long been popular in Brussels, Belgium, and may have originated there.

Figure: *Brussels sprouts*

Cultivation

Forerunners to modern Brussels sprouts were likely cultivated in ancient Rome. Brussels sprouts as we now know them were grown possibly as early as the 13th century in what is now Belgium. The first written reference dates to 1587. During the 16th century, they enjoyed a popularity in the southern Netherlands that eventually spread throughout the cooler parts of Northern Europe.

Brussels sprouts grow in temperature ranges of 7–24°C (45–75°F), with highest yields at 15–18°C (59–64°F). Fields are ready for harvest 90 to 180 days after planting. The edible sprouts grow like buds in helical patterns along the side of long, thick stalks of about 60 to 120 cm (24 to 47 in) in height, maturing over several weeks from the lower to the upper part of the stalk.

Sprouts may be picked by hand into baskets, in which case several harvests are made of five to 15 sprouts at a time, or by cutting the entire stalk at once for processing, or by mechanical harvester, depending on variety. Each stalk can produce 1.1 to 1.4 kg (2.4 to 3.1 lb), although the commercial yield is about 900 g (2.0 lb) per stalk. Harvest season in temperate zones of the northern latitudes is September to March, making Brussels sprout a traditional winter stock vegetable. In the home garden, harvest can be delayed as quality doesn't suffer from freezing. Sprouts are considered to be sweetest after a frost.

Brussels sprouts are a cultivar of the same species as cabbage, in the same family as collard greens, broccoli, kale, and kohlrabi; they are cruciferous (they belong to the Brassicaceae family; old name Cruciferae). They contain good amounts of vitamin A, vitamin C, folic acid and dietary fibre. Moreover, they are believed to protect against colon cancer, because they contain sinigrin. Although they contain compounds such as goitrin that can act as goitrogens and interfere with thyroid hormone production, realistic amounts in the diet do not seem to have any effect on the function of the thyroid gland in humans.

Breeding research conducted by Syngenta in the Netherlands focusing on compounds known as glucosinolates found in Brussels sprouts has resulted in reduced bitterness and in improved health attributes. These improvements in reducing unpleasant taste through scientific breeding advances have been credited with spurring a "renaissance" and growth in production and consumption of Brussels sprouts hybrids.

Europe

In Continental Europe, the largest producers are the Netherlands, at 82,000 metric tons, and Germany, at 10,000 tons. The United Kingdom has production comparable to that of the Netherlands, but it is not generally exported.

North America

Production of Brussels sprouts in the United States began in the 18th century, when French settlers brought them to Louisiana. Thomas Jefferson grew them at Monticello. The first plantings in California's Central Coast began in the 1920s, with significant production beginning in the 1940s. Currently, several thousand acres are planted in coastal areas of San Mateo, Santa Cruz, and Monterey counties of California, which offer an ideal combination of coastal fog and cool temperatures year-round. The harvest season lasts from June through January. They are also grown in Baja California, Mexico, where the harvest season is from December through June.

Most of the United States production is in California, with a smaller percentage of the crop grown in Skagit Valley, Washington, where cool springs, mild summers and rich soil abounds, and to a lesser degree on Long Island, New York. Total United States production is approximately 32,000 tons, with a value of $27 million. Ontario, Canada produces about 1,000 tons per year.

About 80% to 85% of US production is for the frozen food market, with the remainder for fresh consumption. Once harvested, sprouts last three to five weeks under ideal near-freezing conditions before wilting and discolouring, and about half as long at refrigerator temperature. American varieties are generally 2.5–5 cm (0.98–1.97 in) in diametre.

Nutritional and Medicinal Value

Brussels sprouts, as with broccoli and other brassicas, contains sulforaphane, a chemical believed to have potent anticancer properties. Although boiling reduces the level of the anticancer compounds, steaming and stir frying do not result in significant loss. Brussels sprouts and other brassicas are also a source of indole-3-carbinol, a chemical which boosts DNA repair in cells and appears to block the growth of cancer cells.

Consuming Brussels sprouts in excess may not be suitable for heart patients taking anticoagulants since they contain vitamin K. In one such reported incident, doctors determined that the reason for a heart patient's worsening condition was eating too many Brussels sprouts.

Cooking and Preparation

The most common method of preparing Brussels sprouts for cooking begins with cutting the buds off the stalk. Any surplus stem is cut away, and any loose surface leaves are peeled and discarded. Once cut and cleaned, the buds are typically cooked by boiling, steaming, stir frying, grilling, or roasting; however, boiling results in significant loss of anticancer compounds. To ensure even cooking throughout, buds of a similar size are usually chosen. Some cooks will make a single cut or a cross in the centre of the stem to aid the penetration of heat. Brussels sprouts can be pickled as an alternative to cooking.

Overcooking will render the buds gray and soft, and they then develop a strong flavour and odour that some dislike. The odour is associated with glucosinolate sinigrin, an organic compound that contains sulfur: hence the strong smell.

For taste, roasting Brussels sprouts is a common way to cook them in large quantities that seems to bring out the flavour that school children can enjoy. One school district served roasted and pickled Brussels sprouts to 20,000 children who reportedly enjoyed the food during a single day.

Common toppings or additions for Brussels sprouts include parmesan cheese and butter, balsamic vinegar, apple cider vinegar, bacon, pistachios, pine nuts, mustard, brown sugar, and pepper.

A popular way of cooking Brussels sprouts is to sauté them.

Cabbage

Cabbage (*Brassica oleracea* or variants) is a leafy green or purple biennial plant, grown as an annual vegetable crop for its dense-leaved heads. Closely related to other cole crops, such as broccoli, cauliflower, and brussels sprouts, it descends from *B. oleracea var. oleracea*, a wild field cabbage. Cabbage heads generally range from 0.5 to 4 kilograms (1 to 9 lb), and can be green, purple and white. Smooth-leafed firm-headed green cabbages are the most common, with smooth-leafed red and crinkle-leafed savoy cabbages of both colours seen more rarely. It is a multi-layered vegetable. Under conditions of long sunlit days such as are found at high northern latitudes in summer, cabbages can grow much larger. Some records are discussed at the end of the history section.

It is difficult to trace the exact history of cabbage, but it was most likely domesticated somewhere in Europe before 1000 BC, although savoys were not developed until the 16th century. By the Middle Ages, it had become a prominent part of European cuisine. Cabbage heads are generally picked during the first year of the plants' life cycles, but those intended for seed are allowed to grow a second year, and must be kept separated from other cole crops to prevent cross-pollination. Cabbage is prone to several nutrient deficiencies, as well as multiple pests, bacteria and fungal diseases.

The Food and Agriculture Organization of the United Nations (FAO) reports that world production of cabbage and other brassicas for 2011 was almost 69 million metric tons (68 million long tons; 75 million short tons). Almost half of these crops were grown in China, where Chinese cabbage is the most popular *Brassica* vegetable. Cabbages are prepared in many different ways for eating. They can be pickled, fermented for dishes such as sauerkraut, steamed, stewed, sautéed, braised, or eaten raw. Cabbage is a good source of vitamin K, vitamin C and dietary fibre. Contaminated cabbage has been linked to cases of food-borne illness in humans.

Taxonomy and Etymology

Cabbage (*Brassica oleracea* or *B. oleracea* var. *capitata*, var. *tuba*, var. *sabauda* or var. *acephala*) is a member of the genus *Brassica* and the mustard family, Brassicaceae. Several other cruciferous vegetables (sometimes known as *cole crops*) are considered cultivars of *B. oleracea*, including broccoli, collard greens, brussels sprouts, kohlrabi and sprouting broccoli. All of these developed from the wild cabbage *B. oleracea* var. *oleracea*, also called colewort or field cabbage. This original species evolved over thousands of years into those seen today, as selection resulted in cultivars having different characteristics, such as large heads for cabbage, large leaves for kale and thick stems with flower buds for broccoli. The varietal epithet *capitata* is derived from the Latin word for "having a head". *B. oleracea* and its derivatives have hundreds of common names throughout the world.

"Cabbage" was originally used to refer to multiple forms of *B. oleracea*, including those with loose or non-existent heads. A related species, *Brassica rapa*, is commonly named Chinese, napa or celery cabbage, and has many of the same uses. It is also a part of common names for several unrelated species. These include cabbage bark or cabbage tree (a member of the genus *Andira*) and cabbage palms, which include several genera of palms such as *Mauritia*, *Roystonea oleracea*, *Acrocomia* and *Euterpe oenocarpus*.

The original family name of brassicas was *Cruciferae*, which derived from the flower petal pattern thought by medieval Europeans to resemble a crucifix. The word *brassica* derives from *bresic*, a Celtic word for cabbage. Many European and Asiatic names for cabbage are derived from the Celto-Slavic root *cap* or *kap*, meaning "head". The late Middle English word *cabbage* derives from the word *caboche* ("head"), from the Picard dialect of Old French.

This in turn is a variant of the Old French *caboce*. Through the centuries, "cabbage" and its derivatives have been used as slang for numerous items, occupations and activities. Cash and tobacco have both been described by the slang "cabbage", while "cabbage-head" means a fool or stupid person and "cabbaged" means to be exhausted or, vulgarly, in a vegetative state.

Description

Cabbage seedlings have a thin taproot and cordate (heart-shaped) cotyledons. The first leaves produced are ovate (egg-shaped) with a lobed petiole. Plants are 40–60 cm (16–24 in) tall in their first year at the mature vegetative stage, and 1.5–2.0 m (4.9–6.6 ft) tall when flowering in the second year. Heads average between 1 and 8 pounds (0.5 and 4 kg), with fast-growing, earlier-maturing varieties producing smaller heads.

Most cabbages have thick, alternating leaves, with margins that range from wavy or lobed to highly dissected; some varieties have a waxy bloom on the leaves. Plants have root systems that are fibrous and shallow. About 90 percent of the root mass is in the upper 20–30 cm (8–12 in) of soil; some lateral roots can penetrate up to 2 m (6.6 ft) deep.

Figure: *The cabbage inflorescence, which appears in the plant's second year of growth, features white or yellow flowers, each with four perpendicularly arranged petals.*

The inflorescence is an unbranched and indeterminate terminal raceme measuring 50–100 cm (20–40 in) tall, with flowers that are yellow or white. Each flower has four petals set in a perpendicular pattern, as well as four sepals, six stamens, and a superior ovary that is two-celled and contains a single stigma and style. Two of the six stamens have shorter filaments. The fruit is a silique that opens at maturity through dehiscence to reveal brown or black seeds that are small and round in shape. Self-pollination is impossible, and plants are cross-pollinated by insects. The initial leaves form a rosette shape comprising 7 to 15 leaves, each measuring 25–35 cm (10–14 in) by 20–30 cm (8–12 in); after this, leaves with shorter petioles develop and heads form through the leaves cupping inward.

Many shapes, colours and leaf textures are found in various cultivated varieties of cabbage. Leaf types are generally divided between crinkled-leaf, loose-head savoys and smooth-leaf firm-head cabbages, while the colour spectrum includes white and a range of greens and purples. Oblate, round and pointed shapes are found.

Cabbage has been selectively bred for head weight and morphological characteristics, frost hardiness, fast growth and storage ability. The appearance of the cabbage head has been given importance in selective breeding, with varieties being chosen for shape, colour, firmness and other physical characteristics. Breeding objectives are now focused on increasing resistance to various insects and diseases and improving the nutritional content of cabbage. Scientific research into the genetic modification of *B. oleracea* crops, including cabbage, has included European Union and United States explorations of greater insect and herbicide resistance. Genetically modified *B. oleracea* crops are not currently used in commercial agriculture.

Figure: *Green and purple cabbages*

Although cabbage has an extensive history, it is difficult to trace its exact origins owing to the many varieties of leafy greens classified as "brassicas". The wild ancestor of cabbage, *Brassica oleracea*, originally found in Britain and continental Europe, is tolerant of salt but not encroachment by other plants and consequently inhabits rocky cliffs in cool damp coastal habitats, retaining water and nutrients in its slightly thickened, turgid leaves. According to the triangle of U theory of the evolution and relationships between *Brassica* species, *B. oleracea* and other closely related kale vegetables (cabbages, kale, broccoli, Brussels sprouts, and cauliflower) represent one of three ancestral lines from which all other brassicas originated.

Cabbage was probably domesticated later in history than Near Eastern crops such as lentils and summer wheat. Because of the wide range of crops developed from the wild *B. oleracea*, multiple broadly contemporaneous domestications of cabbage may have occurred throughout Europe. Nonheading cabbages and kale were probably the first to be domesticated, before 1000 BC, by the Celts of central and western Europe.

Unidentified brassicas were part of the highly conservative unchanging Mesopotamian garden repertory.

It is believed that the ancient Egyptians did not cultivate cabbage, which is not native to the Nile valley, though a word *shaw't* in Papyrus Harris of the time of Ramesses III, has been interpreted as "cabbage". Ptolemaic Egyptians knew the cole crops as *gramb*, under the influence of Greek *krambe*, which had been a familiar plant to the Macedonian antecedents of the Ptolemies; By early Roman times Egyptian artisans and children were eating cabbage and turnips among a wide variety of other pulses and vegetables.

Greeks had some varieties of cabbage, mentioned by Theophrastus although whether it was more closely related to today's cabbage or to one of the other *Brassica* crops is unknown. According to another view, the headed cabbage variety was known to Greeks as *krambe* and to Romans as *brassica* or *olus*; the open, leafy variety (kale) was known in Greek as *raphanos* and in Latin as *caulis*. In the fourth century Diogenes ate nothing but cabbage and drank nothing but water, and Chrysippus of Cnidos wrote a treatise on cabbage, which Pliny knew, but has not survived.

The Greeks were convinced that cabbages and grapevines were inimical, and that cabbage planted too near the vine would impart its unwelcome odour to the grapes; this Mediterranean sense of antipathy survives today. *Brassica* was considered by some Romans a table luxury, but Lucullus considered it unfit for the senatorial table. More traditionalist Cato the Elder, espousing the simple, Republican life, ate his cabbage cooked or raw, dressed with vinegar; he said it surpassed all other vegetables, gave directions for its medicinal use, which extended to the cabbage-eater's urine, in which infants might be rinsed and approvingly distinguished three varieties. Pliny the Elder listed seven, such as Pompeii cabbage, Cumae cabbage and Sabellian cabbage. According to Pliny, the Pompeii cabbage, which could not stand cold, is "taller, and has a thick stock near the root, but grows thicker between the leaves, these being scantier and narrower, but their tenderness is a valuable quality". The Pompeii cabbage was also mentioned by Columella in *De Re Rustica*. Apicius gives several recipes for *cauliculi*, tender cabbage shoots.

The Greeks and Romans claimed medicinal usages for their cabbage variety that included relief from gout, headaches and the symptoms of poisonous mushroom ingestion. The antipathy towards the vine made it seem that eating cabbage would avoid drunkenness. Cabbage continued to figure in the *materia medica* of Antiquity as well as at table: in the first century AD Dioscorides mentions two kinds of coleworts with medical uses, the cultivated and the wild, and his opinions continued to be paraphrased in herbals right through the 17th century.

At the end of Antiquity cabbage is mentioned in *De observatione ciborum* ("On the Observance of Foods") of Anthemis, a Greek doctor at the court of Theodoric the Great, and cabbage appears among vegetables directed to be cultivated in the *Capitulare de villis*, composed c. 771-800 that guided the governance of the royal estates of Charlemagne.

In Britain the Anglo-Saxon cultivated *cawel*. When round-headed cabbages appeared in 14th-century England they were called *cabaches* and *caboches*, words drawn from Old French and applied at first to refer to the ball of unopened leaves, the contemporaneous recipe that commences "Take cabbages and quarter them, and seethe them in good broth", also suggests the tightly headed cabbage.

Manuscript illuminations show the prominence of cabbage in the cuisine of the High Middle Ages, and cabbage seeds feature among the seed list of purchases for the use of King John II of France when captive in England in 1360, but cabbages were also a familiar staple of the poor: in the lean year of 1420 the "Bourgeois of Paris" noted that "poor people ate no bread, nothing but cabbages and turnips and such dishes, without any bread or salt". French naturalist Jean Ruel made what is considered the first explicit mention of head cabbage in his 1536 botanical treatise *De Natura Stirpium*, referring to it as *capucos coles* ("head-coles"), Sir Anthony Ashley, 1st Baronet, did not disdain to have a cabbage at the foot of his monument in Wimborne St Giles.

In Istanbul Sultan Selim III penned a tongue-in-cheek ode to cabbage: without cabbage the halva feast was not complete. Cabbages spread from Europe into Mesopotamia and Egypt as a winter vegetable, and later followed trade routes throughout Asia and the Americas. The absence of Sanskrit or other ancient Eastern language names for cabbage suggests that it was introduced to South Asia relatively recently. In India, cabbage was one of several vegetable crops introduced by colonizing traders from Portugal, who established trade routes from the 14th to 17th centuries. Carl Peter Thunberg reported that cabbage was not yet known in Japan in 1775.

Many cabbage varieties—including some still commonly grown—were introduced in Germany, France, and the Low Countries. During the 16th century, German gardeners developed the savoy cabbage. During the 17th and 18th centuries, cabbage was a food staple in such countries as Germany, England, Ireland and Russia, and pickled cabbage was frequently seen. Sauerkraut was used by Dutch sailors to prevent scurvy during long ship voyages.

Jacques Cartier first brought cabbage to the Americas in 1541–42, and it was probably planted by the early English colonists, despite the lack of written evidence of its existence there until the mid-17th century. By the 18th century, it was commonly planted by both colonists and native American Indians. Cabbage seeds travelled to Australia in 1788 with the First Fleet, and were planted the same year on Norfolk Island. It became a favourite vegetable of Australians by the 1830s and was frequently seen at the Sydney Markets.

There are several *Guinness Book of World Records* entries related to cabbage. These include the heaviest cabbage, at 57.61 kilograms (127.0 lb), heaviest red cabbage, at 19.05 kilograms (42.0 lb), longest cabbage roll, at 15.37 metres (50.4 ft), and the largest cabbage dish, at 925.4 kilograms (2,040 lb). In 2012, Scott Robb of Palmer, Alaska, broke the world record for heaviest cabbage at 62.71 kilograms (138.25 lb).

Cultivation

Cabbage is generally grown for its densely leaved heads, produced during the first year of its biennial cycle. Plants perform best when grown in well-drained soil in a location that receives full sun. Different varieties prefer different soil types, ranging from lighter sand to heavier clay, but all prefer fertile ground with a pH between 6.0 and 6.8. For optimal growth, there must be adequate levels of nitrogen in the soil, especially during the early head formation stage, and sufficient phosphorus and potassium during the early stages of expansion of the outer leaves. Temperatures between 4 and 24 °C (39 and 75 °F) prompt the best growth, and extended periods of higher or lower temperatures may result in premature bolting (flowering). Flowering induced by periods of low temperatures (a process called vernalization) only occurs if the plant is past the juvenile period. The transition from a juvenile to adult state happens when the stem diametre is about 6 mm (0.24 in). Vernalization allows the plant to grow to an adequate size before flowering. In certain climates, cabbage can be planted at the beginning of the cold period and survive until a later warm period without being induced to flower, a practice that was common in the eastern US.

Plants are generally started in protected locations early in the growing season before being transplanted outside, although some are seeded directly into the ground from which they will be harvested. Seedlings typically emerge in about 4–6 days from seeds planted 1.3 cm (0.5 in) deep at a soil temperature between 20 and 30 °C (68 and 86 °F). Growers normally place plants 30 to 61 cm (12 to 24 in) apart. Closer spacing reduces the resources available to each plant (especially the amount of light) and increases the time taken to reach maturity. Some varieties of cabbage have been developed for ornamental use; these are generally called "flowering cabbage". They do not produce heads and feature purple or green outer leaves surrounding an inner grouping of smaller leaves in white, red, or pink. Early varieties of cabbage take about 70 days from planting to reach maturity, while late varieties take about 120 days. Cabbages are mature when they are firm and solid to the touch. They are harvested by cutting the stalk just below the bottom leaves with a blade. The outer leaves are trimmed, and any diseased, damaged, or necrotic leaves are removed. Delays in harvest can result in the head splitting as a result of expansion of the inner leaves and continued stem growth. Factors that contribute to reduced head weight include: growth in the compacted soils that result from no-till farming practices, drought, waterlogging, insect and disease incidence, and shading and nutrient stress caused by weeds.

When being grown for seed, cabbages must be isolated from other *B. oleracea* subspecies, including the wild varieties, by 0.8 to 1.6 km (0.5 to 1 mi) to prevent cross-pollination. Other *Brassica* species, such as *B. rapa*, *B. juncea*, *B. nigra*, *B. napus* and *Raphanus sativus*, do not readily cross-pollinate.

Cultivars

Figure: *White cabbage*

There are several cultivar groups of cabbage, each including many cultivars:

- Savoy – Characterized by crimped or curly leaves, mild flavour and tender texture
- Spring Greens – Loose-headed, commonly sliced and steamed
- Green – Light to dark green, slightly pointed heads. This is the most commonly grown cultivar.
- Red – Smooth red leaves, often used for pickling or stewing
- White, also called Dutch – Smooth, pale green leaves

Some sources only delineate three cultivars: savoy, red and white, with spring greens and green cabbage being subsumed into the latter.

Cultivation Problems

Due to its high level of nutrient requirements, cabbage is prone to nutrient deficiencies, including boron, calcium, phosphorus and potassium. There are several physiological disorders that can affect the postharvest appearance of cabbage. Internal tip burn occurs when the margins of inside leaves turn brown, but the outer leaves look normal. Necrotic spot is where there are oval sunken spots a few millimetres across that are often grouped around the midrib. In pepper spot, tiny black spots occur on the areas between the veins, which can increase during storage.

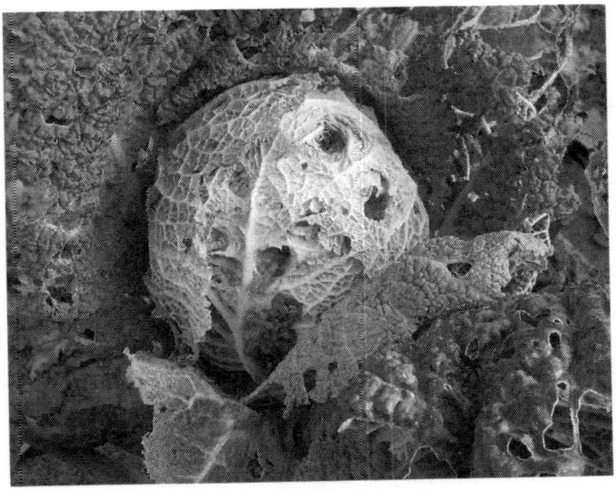

Figure: *Cabbage moth damage to a savoy*

Fungal diseases include wirestem, which causes weak or dying transplants; *Fusarium* yellows, which result in stunted and twisted plants with yellow leaves; and blackleg, which leads to sunken areas on stems and gray-brown spotted leaves. The fungi *Alternaria brassicae* and *A. brassicicola* cause dark leaf spots in affected plants. They are both seedborne and airborne, and typically propagate from spores in infected plant debris left on the soil surface for up to twelve weeks after harvest. *Rhizoctonia solani* causes the post-emergence disease wirestem, resulting in killed seedlings ("damping-off"), root rot or stunted growth and smaller heads.

One of the most common bacterial diseases to affect cabbage is black rot, caused by *Xanthomonas campestris*, which causes chlorotic and necrotic lesions that start at the leaf margins, and wilting of plants. Clubroot, caused by the soilborne slime mold-like organism *Plasmodiophora brassicae*, results in swollen, club-like roots. Downy mildew, a parasitic disease caused by the oomycete *Peronospora parasitica*, produces pale leaves with white, brownish or olive mildew on the lower leaf surfaces; this is often confused with the fungal disease powdery mildew.

Pests include root-knot nematodes and cabbage maggots, which produce stunted and wilted plants with yellow leaves; aphids, which induce stunted plants with curled and yellow leaves; harlequin bugs, which cause white and yellow leaves; thrips, which lead to leaves with white-bronze spots; striped flea beetles, which riddle leaves with small holes; and caterpillars, which leave behind large, ragged holes in leaves. The caterpillar stage of the "small cabbage white butterfly" (*Pieris rapae*), commonly known in the United States as the "imported cabbage worm", is a major cabbage pest in most countries. The large white butterfly (*Pieris brassicae*) is prevalent in eastern European countries. The diamondback moth (*Plutella xylostella*) and the cabbage moth (*Mamestra brassicae*) thrive in the higher summer temperatures of continental Europe, where they cause considerable damage to cabbage crops. In India, the diamondback moth has caused losses up to 90 percent in crops that were not treated with insecticide. Destructive soil insects include the cabbage root fly (*Delia radicum*) and the cabbage maggot (*Hylemya brassicae*), whose larvae can burrow into the part of plant consumed by humans.

Planting near other members of the cabbage family, or where these plants have been placed in previous years, can prompt the spread of pests and disease. Excessive water and excessive heat can also cause cultivation problems.

Production

The Food and Agriculture Organization of the United Nations (FAO) combines cabbage and other brassicas for reporting purposes. Total world production of all brassicas for calendar year 2011 was

68,584,199 metric tons (67,501,016 long tons; 75,601,138 short tons). The nations with the largest production were China, which produced 46 percent of the world total, and India, which produced 12 percent. China and India used a surface area of 958,746 hectares (2,369,110 acres) and 369,000 hectares (910,000 acres), respectively, to grow these crops; the total global surface area used for cabbage and related *Brassica* crops in 2011 was 2,359,103 hectares (5,829,470 acres). The largest yields were from Cyprus, which harvested 118,062.5 kilograms per hectare, Korea (72,122.4 kg/ha), and Ireland (69,635.8 kg/ha).

Cabbages sold for market are generally smaller, and different varieties are used for those sold immediately upon harvest and those stored before sale. Those used for processing, especially sauerkraut, are larger and have a lower percentage of water. Both hand and mechanical harvesting are used, with hand-harvesting generally used for cabbages destined for market sales. In commercial-scale operations, hand-harvested cabbages are trimmed, sorted, and packed directly in the field to increase efficiency. Vacuum cooling rapidly refrigerates the vegetable, allowing for earlier shipping and a fresher product. Cabbage can be stored the longest at –1 to 2 °C (30 to 36 °F) with a humidity of 90–100 percent; these conditions will result in up to six months of longevity. When stored under less ideal conditions, cabbage can still last up to four months.

Culinary Use

Cabbage consumption varies widely around the world: Russia has the highest annual per capita consumption at 20 kilograms (44 lb), followed by Belgium at 4.7 kilograms (10 lb), the Netherlands at 4.0 kilograms (8.8 lb), and Spain at 1.9 kilograms (4.2 lb). Americans consume 3.9 kilograms (8.6 lb) annually per capita.

Cabbage is prepared and consumed in many ways. The simplest options include eating the vegetable raw or steaming it, though many cuisines pickle, stew, sautée or braise cabbage. Pickling is one of the most popular ways of preserving cabbage, creating dishes such as sauerkraut and kimchee, although kimchee is more often made from Chinese cabbage (*B. rapa*). Savoy cabbages are usually used in salads, while smooth-leaf types are utilized for both market sales and processing. Bean curd and cabbage is a staple of Chinese cooking, while the British dish bubble and squeak is made primarily with salt beef and boiled cabbage. In Poland, cabbage is one of the main food crops, and it features prominently in Polish cuisine. Sauerkraut is a frequent dish, either eaten on its own or as a stuffing for other dishes such as golabki (stuffed cabbage) and pierogi (filled pasta). Other eastern European countries, such as Hungary and Romania, also have traditional dishes that feature cabbage as a main ingredient. In India and Ethiopia, cabbage is often included in spicy salads and braises. In the United States, cabbage is used primarily for the production of coleslaw, followed by market use and sauerkraut production.

The characteristic flavour of cabbage is caused by glucosinolates, a class of sulfur-containing glucosides. Although found throughout the plant, these compounds are concentrated in the highest quantities in the seeds; lesser quantities are found in young vegetative tissue, and they decrease as the tissue ages. Cooked cabbage is often criticized for its pungent, unpleasant odour and taste. These develop when cabbage is overcooked and hydrogen sulfide gas is produced.

Nutrition and Health

Cabbage is a good source of vitamin K, vitamin C and fibre. Cabbage is also an excellent source of manganese, vitamin B6, and folate; and a good source of thiamin, riboflavin, calcium, potassium, vitamin A, tryptophan, protein, and magnesium. Studies suggest that it, as well as other cruciferous vegetables, may reduce the risk of some cancers, especially those in the colourectal group. This is possibly due to the glucosinolates found in cole crops, which stimulate the production of detoxifying enzymes that remove carcinogens created during metabolism, or due to the sulphoraphane content, also responsible for metabolic anti-carcinogenic activities. Purple cabbage also contains anthocyanins, which in other vegetables have been shown to have anti-carcinogenic properties. Along with other cole crops, cabbage is a source of indole-3-carbinol, a chemical that boosts DNA repair in cells and has been shown—in experiments using cell cultures and animal models—to block the growth of cancer cells. Research suggests that boiling these vegetables reduces their anti-carcinogenic properties.

Benefits

Glycemic values: Cabbage have extremely low glycemic index (10) and glycemic load (0.58). It is also rich in fibre; soluble fibre lower the GI. Cabbage, at least white, are also a very economic as its cheap and compact making it last for many meals.

Medicinal Lore

In addition to its usual purpose as an edible vegetable, cabbage has been used historically as a medicinal herb for a variety of purported health benefits. The Ancient Greeks recommended consuming the vegetable as a laxative, and used cabbage juice as an antidote for mushroom poisoning, for eye salves, and for liniments used to help bruises heal. In Cato the Elder's work *De Agri Cultura* ("On Agriculture"), he suggested that women could prevent diseases by bathing in urine obtained from those who had frequently eaten cabbage. The ancient Roman nobleman Pliny the Elder described both culinary and medicinal properties of the vegetable, recommending it for drunkenness—both preventatively to counter the effects of alcohol, and to cure hangovers. Similarly, the Ancient Egyptians ate cooked cabbage at the beginning of meals to reduce the intoxicating effects of wine. This traditional usage persisted in European literature until the mid-20th century.

The cooling properties of the leaves were used in Britain as a treatment for trench foot in World War I, and as compresses for ulcers and breast abscesses. Accumulated scientific evidence corroborates that cabbage leaf treatment can reduce the pain and hardness of engorged breasts, and increase the duration of breast feeding. Other medicinal uses recorded in Europe folk medicine include treatments for rheumatism, sore throat, hoarseness, colic, and melancholy. In the United States, cabbage has been used as a hangover cure, to treat abscesses, to prevent sunstroke, or to cool body parts affected by fevers. The leaves have also been used to soothe sore feet and, when tied around the neck of children, to relieve croup. Both mashed cabbage and cabbage juice have been used in poultices to remove boils and treat warts, pneumonia, appendicitis, and ulcers.

Disadvantages

Bloating: Excessive consumption of cabbage may lead to increased intestinal gas which cause bloating and flatulence, due to the trisaccharide raffinose, which the human small intestine cannot break down.

Food-borne Illness

Cabbage has been linked to outbreaks of some food-borne illnesses, including *Listeria monocytogenes* and *Clostridium botulinum*. The latter toxin has been traced to pre-made, packaged coleslaw mixes, while the spores were found on whole cabbages that were otherwise acceptable in appearance. *Shigella* species are able to survive in shredded cabbage. Two outbreaks of *E. coli* in the United States have been linked to cabbage consumption. Biological risk assessments have concluded that there is the potential for further outbreaks linked to uncooked cabbage, due to contamination at many stages of the growing, harvesting and packaging processes. Contaminants from water, humans, animals and soil have the potential to be transferred to cabbage, and from there to the end consumer.

Cauliflower

Cauliflower is one of several vegetables in the species *Brassica oleracea*, in the family Brassicaceae. It is an annual plant that reproduces by seed. Typically, only the head (the *white curd*) is eaten.

The cauliflower head is composed of a white inflorescence meristem. Cauliflower heads resemble those in broccoli, which differs in having flower buds.

Its name is from Latin *caulis* (cabbage) and flower,. *Brassica oleracea* also includes broccoli, brussels sprouts, cabbage, collard greens, and kale, though they are of different cultivar groups.

For such a highly modified plant, cauliflower has a long history. François Pierre La Varenne employed *chouxfleurs* in *Le cuisinier françois*. They were introduced to France from Genoa in the 16th century, and are featured in Olivier de Serres' *Théatre de l'agriculture* (1600), as *cauli-fiori* "as the Italians call it, which are still rather rare in France; they hold an honourable place in the garden because of their delicacy", but they did not commonly appear on grand tables until the time of Louis XIV.

Classification and Identification

Major groups: There are four major groups of cauliflower.

Italian: Diverse in appearance, and biennial and annual in type, this group includes white, Romanesco, various brown, green, purple, and yellow cultivars. This type is the ancestral form from which the others were derived.

Northern European annuals: Used in Europe and North America for summer and fall harvest, it was developed in Germany in the 18th century, and includes the old cultivars Erfurt and Snowball.

Northwest European biennial: Used in Europe for winter and early spring harvest, this was developed in France in the 19th century, and includes the old cultivars Angers and Roscoff.

Asian: A tropical cauliflower used in China and India, it was developed in India during the 19th century from the now-abandoned Cornish type, and includes old varieties Early Benaras and Early Patna.

Varieties

There are hundreds of historic and current commercial varieties used around the world. A comprehensive list of about 80 North American varieties is maintained at North Carolina State University.

Colours

White: White cauliflower is the most common colour of cauliflower.

Orange: Orange cauliflower (*B. oleracea* L. var. *botrytis*) contains 25% more vitamin A than white varieties. This trait came from a natural mutant found in a cauliflower field in Canada. Cultivars include 'Cheddar' and 'Orange Bouquet'.

Green: Green cauliflower, of the *B. oleracea* botrytis group, is sometimes called broccoflower. It is available both with the normal curd shape and a variant spiky curd called Romanesco broccoli.

Both types have been commercially available in the U.S. and Europe since the early 1990s. Green-curded varieties include 'Alverda', 'Green Goddess' and 'Vorda'. Romanesco varieties include 'Minaret' and 'Veronica'.

Purple: The purple colour in this cauliflower is caused by the presence of the antioxidant group anthocyanins, which can also be found in red cabbage and red wine.

Varieties include 'Graffiti' and 'Purple Cape'. In Great Britain and southern Italy, a broccoli with tiny flower buds is sold as a vegetable under the name "purple cauliflower". It is not the same as standard cauliflower with a purple curd.

Nutrition

Cauliflower is low in fat, low in carbohydrates but high in dietary fibre, folate, water, and vitamin C, possessing a high nutritional density.

Cauliflower contains several phytochemicals, common in the cabbage family, that may be beneficial to human health.

- Sulforaphane, a compound released when cauliflower is chopped or chewed, may protect against cancer.
- Other glucosinolates
- Carotenoids
- Indole-3-carbinol, a chemical that enhances DNA repair, and acts as an estrogen antagonist, slowing the growth of cancer cells.

Boiling reduces the levels of these compounds, with losses of 20–30% after five minutes, 40–50% after ten minutes, and 75% after thirty minutes. However, other preparation methods, such as steaming, microwaving, and stir frying, have no significant effect on the compounds.

A high intake of cauliflower has been associated with reduced risk of aggressive prostate cancer.

Cooking

Figure: Aloo gobi, *an Indian dish prepared with cauliflower and potato*

Cauliflower can be roasted, boiled, fried, steamed, or eaten raw. Steaming or microwaving better preserves anticancer compounds than boiling. When cooking, the outer leaves and thick stalks are removed, leaving only the florets. The leaves are also edible, but are most often discarded. The florets should be broken into similar-sized pieces so they are cooked evenly. After eight minutes of steaming, or five minutes of boiling, the florets should be soft, but not mushy (depending on size). Stirring while cooking can break the florets into smaller, uneven pieces.

Low carbohydrate dieters can use cauliflower as a reasonable substitute for potatoes or rice; while they can produce a similar texture, or mouth feel, they lack the starch of the originals.

Fractal Dimension

Cauliflower has been noticed by mathematicians for its distinct fractal dimension, predicted to be about 2.8.

The first reliable reference to cauliflower is found in the writings of the Arab Muslim scientists Ibn al-'Awwam and Ibn al-Baitar, in the 12th and 13th centuries.

Insect Pests of Cauliflower and their Management

Cauliflower (Brassica oleracea L. var. botrytis) is an important winter vegetable grown in India. It belongs to the family Cruciferae. India produces about 4.694 M mt of Cauliflower per year from 0.256 M ha area with an average productivity of 18.3 mt/ha. Bihar, Uttar Pradesh, West Bengal, Orissa, Assam, Haryana, Rajasthan and Maharastra are major cauliflower growing states. It also cultivated in non traditional areas of Tamil Nadu, Andhra Pradesh and Kerala. It is consumed as vegetable in curries, soups and pickles and low in fat and rich source of dietary fibres. In India, the major constraints for production loss are due to physiological disorders, pests and diseases. Off these, insect pests are prime important as it cause serious economic damage to this crop. The pests such as Diamond Back Moth (DBM), Plutella xylostella, Tobacco caterpillar Spodoptera litura, Leaf webber Crocidolomia binotalis, stem borer Hellula undalis, Aphid, Brevicornea brassicae, Mustard aphid, Lipaphis erisimi and Painted bugs, Bagrada cruciferarum are the important pests of cauliflower.

Among these, DBM and tobacco caterpillar are notorious and chemical insecticides are widely used by the farmers to control these pests. Sole relay on synthetic chemicals is not recommended while considering pesticide residues as well other issues like insecticide resistance, resurgence and secondary pests out break. Keeping the facts in view, here we discussed some the integrated approaches for the sustainable pest management programme and it would be highly useful for cauliflower growers across the country.

Diamond Back Moth, Plutella xylostella

It is one of the serious pests of cauliflower in India and worldwide. The grayish brown moth having a typical whit patches, which form diamond like appearance in the wings while in rest. The eggs are laid singly on under surface of leaves veins. The larva is greenish with hairs on the body and it pupates by making a silken web.

Symptoms and Damage

The larva is small and green in colour. It feed on leaves and makes bite holes, causes excessive defoliation, mostly on the growing part i.e primordial. Severe attacks could abort head formation. It may cause yield up to 80 to 90 per cent especially hot periods.

Management

- Planting Indian mustard as trap crop. Sowing of two rows of bold seeded mustard for every 25 rows of cauliflower in twice per season i.e 12 days preceding planting and 40 days later of first planting
- Application of 4 % Neem Seed Kernel Extract at head initiation stage i.e 17 to 28 days after planting
- Repeat of spray of NSKE 4% at 10 to 15 days interval, when the population buildup is high and avoid the spray after curd formation because it could affect curd quality.
- Spray of contact insecticides like quinalphos or chlorpyriphos 0.05%
- Bt formulations also can be used

Leaf Webber, Crocidolomia Binotalis

Moth with brownish fore wing lay overlapping flat eggs in group on the leaves. The larva with red head and body will have the longitudinal strips with short hairs and the pupation takes place in the soil with earthen cocoons.

Symptoms and Damage

Group of hairy larvae damage the leaves. The leaves are skeletonized and larvae remain on the under surface of the leaves in webs. They also damage the heads and in severe case it leads to rotting.

Management

The same management measures recommended for DBM would be followed against this pest also

Stem Borer, Hellula Undalis

Larva looks brownish with black head and four longitudinal strips on the body. Moth appears with wavy lines and central elliptical markings on the fore wings. The pupation takes place inside the burrow of the stem as well in the soil.

Symptoms and Damage

Larvae web the leaves and bore into main stem immediately after transplanting and prevent head formation, which results in production of multiple heads. Larvae also feeds on stalk or leaf veins and in severe attack it makes the produce to unfit for marketing.

Management

- Sowing of mustard 50-60 plant per metre and it needs thick foliage of mustard to reduce the infestation
- Other measures are as like against leaf webber and DBM

Tobacco Caterpillar, Spodoptera Litura

Grey colour moth having the forewings with white strip lines. Larvae are dark brown in colour and gregarious nature.

Symptoms and Damage

Eggs are laid in clusters on under surface of leaves. Young larvae feed gregariously and skeletonize the leaves. Large larvae bore into heads. The infestation starts in the nursery itself so one should need to take preventive measures while preparing the nursery beds to avoid the main field infestation.

Management

- Treat the nursery soil with neem cake at 1.0 kg/m^2
- Soil solarisation with use of white polythene
- Funnel trap baited with pheromone for S. litura @ 12/ha
- Collection and destruction of egg masses and gregariously feeding early instar larvae
- Need based application of SlNPV
- Spray of NSKE 4 % in early stage of larvae
- Collection and destruction of large larvae and use of sharp iron needle to kill the hidden larvae in the curd.

Aphids, Brevicornea Brassicae, and Lipaphis Erisimi

They suck the plant sap from the under surface of the leaves and infested leaves shows whitey mealy covering. Severe cases it affects quality of curd.

Management:

- Planting mustard as trap crop
- Use of NSKE 4% when control is necessary
- Application of Oxydemeton methyl 0.02%

Painted Bug

Bugs are small, black in colour with red and yellow lines. It lays barrel shaped brown eggs.

Symptoms and Damage

Both adults and nymphs suck sap from the leaves, shoots and affect the vigour of the plant, results leaf withers.

Management

Single or spot application of dimethoate and oxydemeton methyl 0.05% would be effective.

4

Salad Vegetables

Coriender

Coriander (*Coriandrum sativum*), also known as cilantro, Chinese parsley or dhania, is an annual herb in the family Apiaceae. Coriander is native to regions spanning from southern Europe and North Africa to southwestern Asia. It is a soft plant growing to 50 cm (20 in) tall. The leaves are variable in shape, broadly lobed at the base of the plant, and slender and feathery higher on the flowering stems. The flowers are borne in small umbels, white or very pale pink, asymmetrical, with the petals pointing away from the centre of the umbel longer (5–6 mm or 0.20–0.24 in) than those pointing towards it (only 1–3 mm or 0.039–0.118 in long). The fruit is a globular, dry schizocarp 3–5 mm (0.12–0.20 in) in diametre. Although sometimes eaten alone, the seeds often are used as a spice or an added ingredient in other foods.

All parts of the plant are edible, but the fresh leaves and the dried seeds are the parts most traditionally used in cooking. Coriander is common in South Asian, Southeast Asian, Indian, Middle Eastern, Caucasian, Central Asian, Mediterranean, Tex-Mex, Latin American, Portuguese, Chinese, African, and Scandinavian cuisine.

Leaves

The leaves are variously referred to as coriander leaves, fresh coriander, Chinese parsley, or (in North America) cilantro. It should not be confused with culantro (*Eryngium foetidum* L.) which is a close relative to coriander (*Coriandrum sativum* L.) but has a distinctly different appearance, a much more potent volatile leaf oil and a stronger smell.

Figure: *Coriander leaves*

The leaves have a different taste from the seeds, with citrus overtones. However, many people experience an unpleasant soapy taste or a rank smell and avoid the leaves. The flavours have also been compared to those of the stink bug, and similar chemical groups are involved (aldehydes). The different perceptions of the coriander leaves' taste is likely genetic, with some people having no response to the aromatic chemical that most find pleasant, while simultaneously being sensitive to certain offending unsaturated aldehydes.

The fresh leaves are an ingredient in many Indian foods (such as chutneys and salads), in Chinese and Thai dishes, in Mexican cooking, particularly in salsa and guacamole and as a garnish, and in salads in Russia and other CIS countries. Chopped coriander leaves are a garnish on Indian dishes such as *dal*. As heat diminishes their flavour, coriander leaves are often used raw or added to the dish immediately before serving. In Indian and Central Asian recipes, coriander leaves are used in large amounts and cooked until the flavour diminishes. The leaves spoil quickly when removed from the plant, and lose their aroma when dried or frozen.

Fruits

The dry fruits are known as coriander seeds. In Indian cuisine they are called *dhaṇia*.

The word *coriander* in food preparation may refer solely to these seeds (as a spice), rather than to the plant. The seeds have a lemony citrus flavour when crushed, due to terpenes linalool and pinene. It is described as warm, nutty, spicy, and orange-flavoured.

The nutritional profile of coriander seed is different from the fresh stems and leaves, the vitamin content being less than amounts being displayed in the chart above for the plant, with some being absent entirely. However, the seeds do provide significant amounts of calcium, iron, magnesium, and manganese.

***Figure:** Dried coriander fruits, often called* coriander seeds *when used as a spice*

The variety *C. s. vulgare* has a fruit diametre of 3–5 mm (0.12–0.20 in), while var. *microcarpum* fruits have a diametre of 1.5–3 mm (0.059–0.118 in). Large-fruited types are grown mainly by tropical and subtropical countries, e.g. Morocco, India and Australia, and contain a low volatile oil content (0.1-0.4%). They are used extensively for grinding and blending purposes in the spice trade. Types with smaller fruit are produced in temperate regions and usually have a volatile oil content of around 0.4-1.8%, so are highly valued as a raw material for the preparation of essential oil.

It is commonly found both as whole dried seeds and in ground form. Seeds can be roasted or heated on a dry pan briefly before grinding to enhance and alter the aroma. Ground coriander seed loses flavour quickly in storage and is best ground fresh.

Coriander seed is a spice in *garam masala* and Indian curries, which often employ the ground fruits in generous amounts together with cumin. It acts as a thickener.

Roasted coriander seeds, called *dhana dal*, are eaten as a snack. They are the main ingredient of the two south Indian dishes: *sambhar* and *rasam*. Coriander seeds are boiled with water and drunk as indigenous medicine for colds.

Outside of Asia, coriander seed is used widely in the process for pickling vegetables. In Germany and South Africa the seeds are used while making sausages. In Russia and Central Europe, coriander seed is an occasional ingredient in rye bread (e.g. borodinsky bread), as an alternative to caraway. Coriander seeds are used in European cuisine today, though they were more important in former centuries.

The Zuni people have adapted it into their cuisine, mixing the powdered seeds ground with chile and using it as a condiment with meat, and eating leaves as a salad.

Coriander seeds are used in brewing certain styles of beer, particularly some Belgian wheat beers. The coriander seeds are used with orange peel to add a citrus character.

Roots

Coriander roots have a deeper, more intense flavour than the leaves. They are used in a variety of Asian cuisines. They are commonly used in Thai dishes, including soups and curry pastes.

Coriander grows wild over a wide area of the Near East and southern Europe, prompting the comment, "It is hard to define exactly where this plant is wild and where it only recently established itself." Fifteen desiccated mericarps were found in the Pre-Pottery Neolithic B level of the Nahal Hemel Cave in Israel, which may be the oldest archaeological find of coriander. About half a litre of coriander mericarps were recovered from the tomb of Tutankhamen, and because this plant does not grow wild in Egypt, Zohary and Hopf interpret this find as proof that coriander was cultivated by the ancient Egyptians.

Coriander seems to have been cultivated in Greece since at least the second millennium BC. One of the Linear B tablets recovered from Pylos refers to the species as being cultivated for the manufacture of perfumes, and it appears that it was used in two forms: as a spice for its seeds and as a herb for the flavour of its leaves. This appears to be confirmed by archaeological evidence from the same period: the large quantities of the species retrieved from an Early Bronze Age layer at Sitagroi in Macedonia could point to cultivation of the species at that time.

Coriander was brought to the British colonies in North America in 1670, and was one of the first spices cultivated by early settlers.

Arugula

Eruca sativa (syn. *E. vesicaria* subsp. *sativa* (Miller) Thell., *Brassica eruca* L.) is an edible annual plant, commonly known as salad rocket, roquette, rucola, rugula, colewort, and, in the United States, arugula. It is sometimes conflated with *Diplotaxis tenuifolia*, the perennial wall rocket, another plant of the Brassicaceae family, which in the past was used in the same manner. Eruca sativa, which is widely popular as a salad vegetable, is a species of *Eruca* native to the Mediterranean region, from Morocco and Portugal in the west to Syria, Lebanon and

Turkey in the east. The Latin adjective *sativa* in the plant's binomial is derived from *satum*, the supine of the verb *sero*, meaning "to sow", indicating that the seeds of the plant were sown in gardens. *Eruca sativa* differs from *E. vesicaria* in having early deciduous sepals. Some botanists consider it a subspecies of *Eruca vesicaria*: *E. vesicaria* subsp. *sativa*. Still others do not differentiate between the two.

Other common names include garden rocket, or more simply rocket (British, Australian, Canadian, South African and New Zealand English), and eruca. The English common name, *rocket*, derives from the French *roquette*, a diminutive of the Latin word *eruca*, which designated an unspecified plant in the Brassicaceae family (probably a type of cabbage). Arugula, the common name now widespread in the United States, entered American English from non-standard (dialect) Italian. (The standard Italian word is *rucola*, a diminutive of the Latin "eruca"). The Oxford English Dictionary dates the first appearance of "arugula" in American English to a 1960 *New York Times* article by food editor and prolific cookbook writer, Craig Claiborne,

Eruca sativa grows 20–100 centimetres (8–39 in) in height. The leaves are deeply pinnately lobed with four to ten small lateral lobes and a large terminal lobe. The flowers are 2–4 cm (0.8–1.6 in) in diametre, arranged in a corymb in typical Brassicaceae fashion; with creamy white petals veined with purple, and with yellow stamens; the sepals are shed soon after the flower opens. The fruit is a siliqua (pod) 12–35 millimetres (0.5–1.4 in) long with an apical beak, and containing several seeds (which are edible). The species has a chromosome number of $2n = 22$.

Eruca sativa typically grows on dry, disturbed ground and is also used as a food by the larvae of some moth species, including the Garden Carpet moth. Eruca sativa roots are also susceptible to nematode infestation.

Cultivation and History

A pungent, leafy green vegetable resembling a longer-leaved and open lettuce, rocket is rich in vitamin C and potassium. In addition to the leaves, the flowers, young seed pods and mature seeds are all edible.

Grown as an edible herb in the Mediterranean area since Roman times, it was mentioned by various classical authors as an aphrodisiac, most famously in a poem long ascribed to Virgil, *Moretum*, which contains the line: "*et veneris revocans eruca morantuem*" ("the rocket excites the sexual desire of drowsy people"). Some writers assert that for this reason during the Middle Ages it was forbidden to grow rocket in monasteries. It was listed, however, in a decree by Charlemagne of 802 as one of the pot herbs suitable for growing in gardens. Gillian Reilly, author of the *Oxford Companion to Italian Food,* states that because of its reputation as a sexual stimulant, it was "prudently mixed with lettuce, which was the opposite" (i.e., calming or even soporific). Reilly continues that "nowadays rocket is enjoyed innocently in mixed salads, to which it adds a pleasing pungency".

Figure: *Flower of* Eruca sativa

Rocket was traditionally collected in the wild or grown in home gardens along with such herbs as parsley and basil. It is now grown commercially from the Veneto to Iowa to Brazil, and is available for purchase in supermarkets and farmers' markets throughout the world. It is also naturalised as a wild plant away from its native range in temperate regions around the world, including northern Europe and North America. In India, the mature seeds are known as Gargeer.

Growing rocket in mild frost conditions stymies the growth of the plant, as well as turning the green leaves red.

Uses

Rocket has a pungent, peppery flavour that is exceptionally strong for a leafy green. It is frequently used in salads, often mixed with other greens in a mesclun. It is also used raw with pasta or meats in northern Italy and in western Slovenia (especially in the Slovenian Istria). In Italy, raw rocket is often added to pizzas just before the baking period ends or immediately afterwards, so that it will not wilt in the heat. It is also used cooked in Puglia, in Southern Italy, to make the pasta dish *cavatiéddi*, "in which large amounts of coarsely chopped rocket are added to pasta seasoned with a homemade reduced tomato sauce and pecorino", as well as in "many unpretentious recipes in which it is added, chopped, to sauces and cooked dishes" or in a sauce (made by frying it in olive oil and garlic) used a condiment for cold meats and fish. In the Slovenian Littoral, it is often combined with boiled potatoes, used in a soup, or served with the cheese burek, especially in the town of Koper.

A sweet, peppery digestive alcohol called *rucolino* is made from rocket on the island of Ischia in the Gulf of Naples. This liqueur is a local speciality enjoyed in small quantities following a meal in the same way as a limoncello or grappa.

In Brazil, where its use is widespread, rocket is eaten raw in salads. A popular combination is rocket mixed with mozzarella cheese (normally made out of buffalo milk) and sun-dried tomatoes.

In Egypt the plant is commonly eaten raw as a side dish with many meals, with ful medames for breakfast, and regularly accompanies local seafood dishes.

In West Asia and Northern India, *Eruca* seeds are pressed to make taramira oil, used in pickling and (after aging to remove acridity) as a salad or cooking oil. The seed cake is also used as animal feed.

Carrots

The carrot (*Daucus carota* subsp. *sativus*; etymology: from Late Latin *carôta*, from Greek êáñùôüí *karôton*, originally from the Indo-European root *ker-* (horn), due to its horn-like shape) is a root vegetable, usually orange in colour, though purple, red, white, and yellow varieties exist. It has a crisp texture when fresh. The most commonly eaten part of a carrot is a taproot, although the greens are sometimes eaten as well. It is a domesticated form of the wild carrot *Daucus carota*, native to Europe and southwestern Asia. The domestic carrot has been selectively bred for its greatly enlarged and more palatable, less woody-textured edible taproot. The Food and Agriculture Organization of the

United Nations (FAO) reports that world production of carrots and turnips (these plants are combined by the FAO for reporting purposes) for calendar year 2011 was almost 35.658 million tonnes. Almost half were grown in China. Carrots are widely used in many cuisines, especially in the preparation of salads, and carrot salads are a tradition in many regional cuisines.

The carrot is a variable biennial plant, usually growing up to 1 m tall and flowering from June to August. The umbels are claret-coloured or pale pink before they open, then bright white and rounded when in full flower, measuring 3–7 cm wide with a festoon of bracts beneath; finally, as they turn to seed, they contract and become concave like a bird's nest. The dried umbels detach from the plant, becoming tumbleweeds.

Similar in appearance to the deadly poison hemlock, *D. carota* is distinguished by a mix of bi-pinnate and tri-pinnate leaves, fine hairs on its stems and leaves, a root that smells like carrots, and occasionally a single dark red flower in its centre.

Chemistry

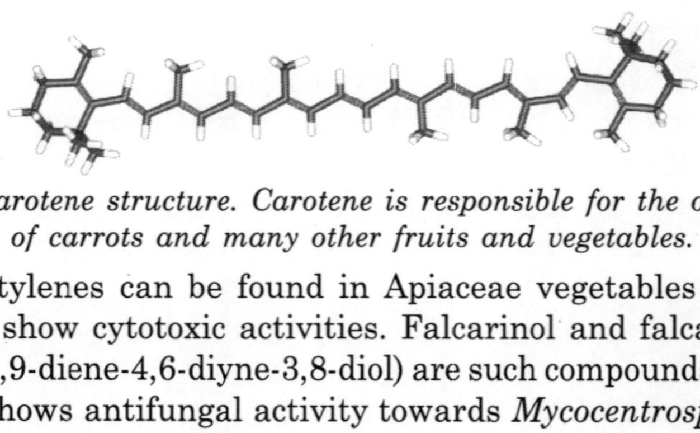

Figure: *β-Carotene structure. Carotene is responsible for the orange colour of carrots and many other fruits and vegetables.*

Polyacetylenes can be found in Apiaceae vegetables like carrots where they show cytotoxic activities. Falcarinol and falcarindiol (*cis*-heptadeca-1,9-diene-4,6-diyne-3,8-diol) are such compounds. This latter compound shows antifungal activity towards *Mycocentrospora acerina* and *Cladosporium cladosporioides*. Falcarindiol is the main compound responsible for bitterness in carrots. Other compounds such as pyrrolidine (present in the leaves), 6-hydroxymellein, 6-methoxymellein, eugenin, 2,4,5-trimethoxybenzaldehyde (gazarin) or (Z)-3-acetoxy-heptadeca-1,9-diene-4,6-diin-8-ol (falcarindiol 3-acetate) can also be found in carrot.

Methods of Consumption and Uses

Carrots can be eaten in a variety of ways. Only 3% of the β-carotene in raw carrots is released during digestion: this can be improved to 39% by pulping, cooking and adding cooking oil. Alternatively they may be chopped and boiled, fried or steamed, and cooked in soups and stews, as well as baby and pet foods. A well-known dish is *carrots julienne*.

Together with onion and celery, carrots are one of the primary vegetables used in a *mirepoix* to make various broths.

Figure: *Carrot tzimmes*

The greens are edible as a leaf vegetable, but are only occasionally eaten by humans; some sources suggest that the greens contain toxic alkaloids. When used for this purpose, they are harvested young in high-density plantings, before significant root development, and typically used stir-fried, or in salads. Some people are allergic to carrots. In a 2010 study on the prevalence of food allergies in Europe, 3.6 percent of young adults showed some degree of sensitivity to carrots. Because the major carrot allergen, the protein Dauc c 1.0104, is cross-reactive with homologues in birch pollen (Bet v 1) and mugwort pollen (Art v 1), most carrot allergy sufferers are also allergic to pollen from these plants.

In India carrots are used in a variety of ways, as salads or as vegetables added to spicy rice or daal dishes. A popular variation in north India is the Gajar Ka Halwa carrot dessert, which has carrots grated and cooked in milk until the whole mixture is solid, after which nuts and butter are added. Carrot salads are usually made with grated carrots in western parts with a seasoning of mustard seeds and green chillies popped in hot oil, while adding carrots to rice usually is in julienne shape.

The variety of carrot found in north India is rare everywhere except in Central Asia and other contiguous regions, and is now growing in popularity in larger cosmopolitan cities in South India. The north Indian carrot is pink-red comparable to plum or raspberry or deep red apple in colour (without a touch of yellow or blue) while most other carrot varieties in the world vary from orange to yellow in colour, comparable to hallowe'en pumpkins. Since the late 1980s, baby carrots or mini-carrots (carrots that have been peeled and cut into uniform cylinders) have been a popular ready-to-eat snack food available in many supermarkets. Carrots are purέed and used as baby food, dehydrated to make chips, flakes, and powder, and thinly sliced and deep-fried, like potato chips.

The sweetness of carrots allows the vegetable to be used in some fruit-like roles. Grated carrots are used in carrot cakes, as well as carrot puddings, an old English dish thought to have originated in the early 19th century. Carrots can also be used alone or with fruits in jam and preserves. Carrot juice is also widely marketed, especially as a health drink, either stand-alone or blended with fruits and other vegetables.

Companion Plant

Carrots are useful companion plants for gardeners. There is experimental evidence that growing it intercropped with tomatoes increases tomato production. If left to flower, it (like any umbellifer) attracts predatory wasps that kill many garden pests.

Cultivation

Carrots grow best in full sun but tolerate some shade. The optimum growth temperature is between 16 and 21 °C (61 and 70 °F). In order to avoid growing deformed carrots it is better to plant them in loose soil free from rocks. They thrive in raised garden beds. High nitrogen levels should be avoided, as this will cause the vegetables to become hairy and misshapen. The seeds, which are 1–3 mm in diametre, should be sown about 2 cm deep. Carrots take around four months to mature and it is suggested that carrot seeds are sown from mid-February to July.

Cultivation Problems

There are several diseases that can reduce the yield and market value of carrots. The most devastating carrot disease is *Alternaria* leaf blight, which has been known to eradicate entire crops. A bacterial leaf blight caused by *Xanthomonas campestris* can also be destructive in warm, humid areas. Root knot nematodes (*Meloidogyne* species) can cause stubby or forked roots, or galls. Cavity spot, caused by the oomycetes *Pythium violae* and *Pythium sulcatum*, results in irregularly shaped, depressed lesions on the taproots.

Physical damage can also reduce the value of carrot crops. The two main forms of damage are splitting, whereby a longitudinal crack develops during growth that can be a few centimetres to the entire length of the root, and breaking, which occurs postharvest. These disorders can affect over 30% of commercial crops. Factors associated with high levels of splitting include wide plant spacing, early sowing, lengthy growth durations, and genotype.Carrot fly

The carrot fly (*Chamaepsila rosae*) is a pest of gardens and farms, and mainly affects the crop of carrots, but can also attack parsnips, parsley and celery. It is a member of the family Psilidae (order Diptera).

Larvae

Crop damage is caused by the creamy-yellow larvae (maggots) feeding on the outer layers of the carrot root. In autumn, they may penetrate further into the root. The legless larvae are up to 10 mm in length.

Symptoms of Infestation

Foliage becomes wilted and discoloured. Leaves turn rusty red to scarlet with some yellowing. Rusty-brown tunnels are seen under the outer skin of mature roots.

Control

The flies lay their eggs around the developing carrots; the larvae, once hatched, burrow into the root. As female carrot flies are very low flying, the best method of prevention is to erect a barrier around the crop at least 60 cm high. Alternatively horticultural fleece may be used as a floating mulch to cover the crop. Some plants such as Rosemary, Sage, and Marigold are used to deter the carrot fly. Newer varieties of carrot which claim to be resistant to carrot fly (e.g. "Flyaway") may be used. Another method of control is to use heavier, fine plastic mesh available from garden centres and over the internet made specially for the purpose. This can be in the form of a surrounding wall or a complete "cage". Some form of support framework is necessary to prevent the netting from flattening the foliage. Also the bottom of the netting needs to be in close contact or buried in the soil as it is believed that the eggs are laid on bare soil. When the maggots hatch they then crawl towards the carrots so a soil level barrier is necessary.

Commercially available beneficial nematodes (*Steinernema* spp.) can be applied to the soil surrounding the carrot crop, where they will infect the damaging carrot root fly larvae.

Cultivars

Carrot cultivars can be grouped into two broad classes, eastern carrots and western carrots. More recently, a number of novelty cultivars have been bred for particular characteristics. The city of Holtville, California, promotes itself as "Carrot Capital of the World", and holds an annual festival devoted entirely to the carrot.

Eastern carrots were domesticated in Central Asia, probably in modern-day Iran and Afghanistan in the 10th century, or possibly earlier. Specimens of the eastern carrot that survive to the present day are commonly purple or yellow, and often have branched roots. The purple colour common in these carrots comes from anthocyanin pigments.

The western carrot emerged in the Netherlands in the 17th century, its orange colour making it popular in those countries as an emblem of the House of Orange and the struggle for Dutch independence. The orange colour results from abundant carotenes in these cultivars.

Western carrot cultivars are commonly classified by their root shape. The four general types listed below are followed by specific variety examples. However, there are also many varieties falling outside these types.

- Chantenay carrots are shorter than other cultivars, but have greater girth, sometimes growing up to 8 centimetres (3 in) in diametre. They have broad shoulders and taper towards a blunt, rounded tip. They are most commonly diced for use in canned or prepared foods.
  - Carson Hybrid
  - Red Cored Chantenay
- Danvers carrots have a conical shape, having well-defined shoulders and tapering to a point at the tip. They are somewhat shorter than Imperator cultivars, but more tolerant of heavy soil. Danvers cultivars are often puréed as baby food. They were developed in 1871 in Danvers, Massachusetts.
  - Danvers Half Long
  - Danvers 126
- Imperator carrots are sweet and generally longer and narrower than other types. Imperator types are the most widely produced by commercial growers.
  - Imperator 58
  - Sugarsnax Hybrid
- Nantes carrots are shorter with a more blunt tip than Imperator types but still can attain high yields in varying conditions. A number of hybrid varieties are available.

- o Nelson Hybrid
- o Scarlet Nantes
- o Sweetness Hybrid

One particular variety lacks the usual orange pigment from carotenes, owing its white colour to a recessive gene for tocopherol (vitamin E).

Production Trends

Carrot is one of the top-ten most economically important vegetables crops in the world. In 2011, according to the Food and Agriculture Organization of the United Nations, 35.658 million tonnes of carrots and turnips were produced worldwide for human consumption, grown on 1,184,000 hectares (2,926,000 acres). With 16.233 million tonnes, China was by far the largest producer and accounted for 45.5% of the global output, followed by Russia (1.735 million tonnes), the United States (1.342), Uzbekistan (1.222), Poland (0.887), Ukraine (0.864), and the United Kingdom (0.694). About 61% of world carrot production occurred in Asia, followed by the Europe (24.2%) and the Americas (North, Central, and South America and the Caribbean) (9.7%). Less than 4% of the world's 2011 total was produced in Africa. Global production has increased from 21.4 million tonnes in 2000, 13.7 million tonnes in 1990, 10.4 million tonnes in 1980, and 7.85 million tonnes in 1970. The rate of increase in the global production of carrots has been greater than the world's population growth rate, and greater than the overall increase in world vegetable production. Europe was traditionally the major centre of production, but was overtaken by Asia in 1997. The growth in global production is largely the result of increases in production area rather than average yield. Modest improvements in the latter can be attributed to optimised agricultural practices, the development of better cultivars (including hybrids), and increased farm mechanisation.

Storage

Carrots can be stored for several months in the refrigerator or over winter in a moist, cool place. For long term storage, unwashed carrots can be placed in a bucket between layers of sand, a 50/50 mix of sand and wood shavings, or in soil. A temperature range of 32 to 40°F (0 to 5°C) is best.

Radish

The radish (*Raphanus sativus*) is an edible root vegetable of the Brassicaceae family that was domesticated in Europe in pre-Roman times. They are grown and consumed throughout the world. Radishes have numerous varieties, varying in size, colour and duration of required cultivation time. There are some radishes that are grown for their seeds; oilseed radishes are grown, as the name implies, for oil production. Radish can sprout from seed to small plant in as little as 3 days.

The descriptive Greek name of the genus *Raphanus* means "quickly appearing" and refers to the rapid germination of these plants. *Raphanistrum*, from the same Greek root, is an old name once used for this genus. The common name "radish" is derived from Latin *radix* (root).

Description of the Radish Plant

Radishes are round to cylindrical with a colour ranging from white to red. A longer root form, ideal for cooking, grows up to 15 cm (6 in) long, while the smaller, rounder form is typically eaten raw in salads. The flesh initially tastes sweet, but becomes bitter if the vegetable is left in the ground for too long. Leaves are arranged in a rosette, with sizes ranging from 10–15 cm (4–6 in) in small cultivars, to up to 45 cm (18 in) in large cultivars. They have a lyrate shape, meaning they are divided pinnately with an enlarged terminal lobe and smaller lateral lobes. The white flowers are borne on a racemose inflorescence.

The radish is a diploid species, and has 18 chromosomes ($2n$=18).

Cultivation

Figure: *Growing radish plants*

Radishes grow best in full sun and light, sandy loams with pH 6.5–7.0. They are in season from April to June and from October to January in most parts of North America; in Europe and Japan they are available year-round due to the plurality of varieties grown.

Summer radishes mature rapidly, with many varieties germinating in 3–7 days, and reaching maturity in three to four weeks. Harvesting periods can be extended through repeated plantings, spaced a week or two apart.

As with other root crops, tilling the soil to loosen it up and remove rocks helps the roots grow. However, radishes are used in no-till farming to help reverse compaction.

Most soil types will work, though sandy loams are particularly good for winter and spring crops, while soils that form a hard crust can impair growth. The depth at which seeds are planted affects the size of the root, from 1 cm (0.4 in) deep recommended for small radishes to 4 cm (1.6 in) for large radishes.

Radishes are a common garden crop in the U.S., and the fast harvest cycle makes them a popular choice for children's gardens.

In temperate climates, it is customary to plant radishes every two weeks from early spring until a few weeks before the first frost, except during periods of hot weather. In warm-weather climates, they are normally planted in the fall.

After harvest, radishes can be stored without loss of quality for two or three days at room temperature, and about two months at 0 °C (32 °F) with a relative humidity of 90–95%.

Companion Plant

Radishes serve as companion plants for many other species, because of their ability to function as a trap crop against pests like flea beetles. These pests will attack the leaves, but the root remains healthy and can be harvested later.

Pests Management

Although often unintentionally introduced, the larvae of *Pieris rapae*, the small white butterfly is known to pest on the leaves of the radish.

Flea Beetles

The term "flea beetle" covers a number of species, all members of the Chrysomelidae family. They are sucking insects that riddle radish leaves and other plants with small holes about 1/16 inch in diametre. The leaf damage typically won't hurt the developing radish root, because radishes grow so quickly – 20 to 25 days average from seed to harvest. However, the pests carry diseases that can infect other plants, and their leaf damage to other plants can cause problems. Flea beetles thrive in warm, dry conditions, and they attack plants that are stressed by lack of water. Irrigation helps your radishes grow as well as helping them survive flea beetle attacks. Several effective organic compounds are reported by Appropriate Transfer Technology for Rural Areas for controlling flea beetles, including neem, rotenone, sabidilla and pyrethrins, plus extract of garlic, onion and mint or interplanting them with your radishes to repel flea beetles.

Thrips

Thrips are categorized in the Thysanoptera order. Many kinds of thrips exist. Some are beneficial, feeding on mites and insects or fungus spores. Pest thrips are sucking insects that puncture leaves to feed on the sap. They don't fly well, even though they are winged, but they can float on winds over great distances. Thrip damage appears as scarred or wrinkled foliage on radishes. If the damage occurs at the seedling stage, it may affect the root development or stunt the plant. Neem oil or pyrethrin products are effective treatments. Green lacewings and minute pirate bugs feed on thrips, and predatory species of thrips help control the plant-sucking thrips. A simple shot of water from the garden hose knocks thrips off the plants.

Radish Maggot

The radish maggot is the larval stage of Delia planipalpis, a tiny fly. The maggots burrow into the soil around radishes, turnips and cole plants and bore into the root, where you may discover them when you harvest your radishes. If the maggot has moved on to adulthood, you may see the telltale remnants of it as a black spot in the radish. Pull up infected plants and dispose of them in the trash. If you put them in the compost pile, the larvae will complete their life cycle to adulthood and repopulate your garden.

Bagrada Bug

First documented in 2008 in Pasadena, the Bagrada bug is now found in Orange, Ventura, Imperial and Riverside counties. The spread of Bagrada hilaris, sometimes called harlequin bug or painted bug, is being monitored closely. It appears in some commercial crop fields, but has been reported mostly in organic crops and community or home gardens. Bagrada bugs are brownish insects about a quarter-inch long with bright white and orange markings that are clearly visible. They are related to stinkbugs and have a similar shield-shape.

Bagrada bugs suck sap from the leaves of plants, including radishes, leaving them wrinkled or stippled, stunting or disfiguring the plant. They lay eggs in the soil, and frequent cultivation helps control them. Rotate crops and keep debris cleaned up to break the life cycle between planting seasons and crops. Dusting or spraying with pyrethrin or rotenone is another management technique.

Varieties

Broadly speaking, radishes can be categorized into four main types (summer, fall, winter, and spring) and a variety of shapes lengths, colours, and sizes, such as red, pink, white, gray-black or yellow radishes, with round or elongated roots that can grow longer than a parsnip.

Spring or Summer Radishes

Sometimes referred to as European radishes or spring radishes if they are planted in cooler weather, summer radishes are generally small and have a relatively short 3–4 week cultivation time.

Cucumbers

Cucumber (*Cucumis sativus*) is a widely cultivated plant in the gourd family Cucurbitaceae. It is a creeping vine that bears cylindrical fruits that are used as culinary vegetables. There are three main varieties of cucumber: *slicing*, *pickling*, and *burpless*. Within these varieties, several different cultivars have emerged. The cucumber is originally from Southern Asia, but now grows on most continents. Many different varieties are traded on the global market.

The cucumber is a creeping vine that roots in the ground and grows up trellises or other supporting frames, wrapping around supports with thin, spiraling tendrils.

The plant has large leaves that form a canopy over the fruit. The fruit of the cucumber is roughly cylindrical, elongated with tapered ends, and may be as large as 60 centimetres (24 in) long and 10 centimetres (3.9 in) in diametre. Having an enclosed seed and developing from a flower, botanically speaking, cucumbers are classified as accessory fruits. Much like tomatoes and squash they are often also perceived, prepared and eaten as vegetables. Cucumbers are usually more than 90% water.

Flowering and Pollination

A few cultivars of cucumber are parthenocarpic, the blossoms creating seedless fruit without pollination. Pollination for these cultivars degrades the quality. In the United States, these are usually grown in greenhouses, where bees are excluded. In Europe, they are grown outdoors in some regions, and bees are excluded from these areas.

Most cucumber cultivars, however, are seeded and require pollination. Thousands of hives of honey bees are annually carried to cucumber fields just before bloom for this purpose. Cucumbers may also be pollinated by bumblebees and several other bee species. Most cucumbers that require pollination are self-incompatible, so pollen from a different plant is required to form seeds and fruit. Some self-compatible cultivars exist that are related to the 'Lemon' cultivar. Symptoms of inadequate pollination include fruit abortion and misshapen fruit. Partially pollinated flowers may develop fruit that are green and develop normally near the stem end, but are pale yellow and withered at the blossom end.

Traditional cultivars produce male blossoms first, then female, in about equivalent numbers. Newer gynoecious hybrid cultivars produce almost all female blossoms.

They may have a pollenizer cultivar interplanted, and the number of beehives per unit area is increased, but temperature changes induce male flowers even on these plants, which may be sufficient for pollination to occur.

Insecticide applications for insect pests must be done very carefully to avoid killing off the insect pollinators.

Genome

In 2009, an international team of researchers announced they had sequenced the cucumber genome.

Production

According to the Food and Agriculture Organization of the United Nations, China produced at least 60% of the global output of cucumbers in 2005, followed at a distance by Turkey, Russia, Iran and the United States.

Cultivation

The cucumber originated in India, where a great many varieties have been observed, from *Cucumis hystrix.* It has been cultivated for at least 3,000 years, and was probably introduced to other parts of Europe by the Greeks or Romans. Records of cucumber cultivation appear in France in the 9th century, England in the 14th century, and in North America by the mid-16th century.

Varieties

In human cultivation, the varieties of cucumbers are classified into three main varieties: "slicing", "pickling", and "burpless".

Slicing

Figure: *Slicing cucumbers*

Cucumbers grown to eat fresh are called slicing cucumbers. They are mainly eaten in the unripe green form, since the ripe yellow form normally becomes bitter and sour. Slicers grown commercially for the North American market are generally longer, smoother, more uniform in colour, and have a much tougher skin. Slicers in other countries are smaller and have a thinner, more delicate skin. Smaller slicing cucumbers can also be pickled.

Pickling

Cucumbers can be pickled for flavour and longer shelf-life. Although any cucumber can be pickled, commercial pickles are made from cucumbers specially bred for uniformity of length-to-diametre ratio and lack of voids in the flesh. Those cucumbers intended for pickling, called picklers, grow to about 7 cm (3 in) to 10 cm (4 in) long and 2.5 cm (1 in) wide. Compared to slicers, picklers tend to be shorter, thicker, less regularly shaped, and have bumpy skin with tiny white or black-dotted spines. They are never waxed. Colour can vary from creamy yellow to pale or dark green. Pickling cucumbers are sometimes sold fresh as "Kirby" or "Liberty" cucumbers.

The pickling process removes or degrades much of the nutrient content, especially that of vitamin C. Pickled cucumbers are soaked in brine or a combination of vinegar and brine, although not vinegar alone, often along with various spices. Pickled cucumbers are called "pickles" in the US or "gherkins" or "wallies" in the UK, the latter name being more common in the north of England, where it refers to the large vinegar-pickled cucumbers commonly sold in fish and chip shops. (Although the gherkin is of the same species as the cucumber, it is of a completely different cultivar.)

Figure: *Pickling cucumbers*

Burpless

Burpless cucumbers are sweeter and have a thinner skin than other varieties of cucumber, and are reputed to be easy to digest and to have a pleasant taste. They can grow as long as 2 feet (0.61 m). They are nearly seedless, and have a delicate skin.

Most commonly grown in greenhouses, these parthenocarpic cucumbers are often found in grocery markets, shrink-wrapped in plastic. They are sometimes marketed as seedless or burpless, because the seeds and skin of other varieties of cucumbers are said to give some people gas.

Management Practices for Avoiding Cucumber Problems

Diseases, pests, and disorders remain a threat to efficient cucumber production and good profits by lowering yields, reducing quality, and making harvests unreliable. The risk of these problems can be reduced by wise management.

- Keep records.
- Select adapted cultivars with resistance to downy mildew, powdery mildew, anthracnose, scab, angular leafspot, and cucumber mosaic.
- Practice a 12-month, noncucurbit crop rotation.
- Select fertile, well drained fields.
- Sample fields annually for pH, nutrients, nematodes, soil insects, and weeds.
- Use certified seed treated with fungicides and insecticides.
- Seed 0.5-1.0-inch deep on raised beds after soils warm to 60° F.
- Place one strong colony of bees per acre after blooms are present.
- Apply a general purpose fungicide shortly after emergence and include an insecticide as needed.
- Diagnose all leafspots and plant abnormalities promptly.
- Use a high pressure (250 + psi) sprayer for fungicide and insecticide applications.
- Irrigate fields to ensure 1.0-inch of water each week.
- Monitor migration of the pickleworm and spray as needed.
- Reapply nitrogen and potassium after leaching rains.
- Pick three times a week.
- Avoid injuries during harvesting and handling.
- Use chlorine in all dump, wash, and handling water.

Poor stands may be caused by cold or wet soils, improper seeding depth or fertilization, residual herbicides, insects, and damping-off caused by several soil inhabiting fungi such as *Pythium spp., Rhizoctonia spp., Fusarium spp.,* and some leafspotting fungi during wet periods. CONTROL: Follow general management practices, replant.

Seedcorn maggot, adult, larva, pupa *(Delia platura)* and injury to seedlings. In the spring adults seek germinating seeds to lay eggs. Maggot injury is associated with cold, wet, organic soils. CONTROL: Apply soil insecticide, replant.

Wireworm larva and adults *(Conoderus vespertinus* and C. *falli)* and injury to small plant. Wireworms are associated with fields previously in small grain or not in row crops. CONTROL: Avoid problem fields and use soil insecticide.

Herbicide injury from residues carried over from previous crop, most often corn. Herbicide injury is fairly common as there are many possibilities for misuse. CONTROL: Keep records of herbicides used in fields; select proper herbicide(s), application method and rates; check equipment calibration; avoid drift; clean equipment properly.

Cold injury is very serious on cucumbers. It is most likely to show up in low areas. CONTROL: Follow general management practices, replant.

Fertilizer injury often occurs with banded fertilizer. Symptoms include poor stands, marginal foliar burning, and browning of roots. CONTROL: Adjust equipment properly, broadcast fertilizer according to soil test reports, reapply fertilizers if needed as determined by soil test or soluble salts. Serpentine leafminer larva, pupa and adults *(Liriomyza spp.)* and injury. Their presence is common but serious injury is usually confined to small plants. CONTROL: Early detection and prompt insecticide sprays.

Cucumber beetle damage. This insect overwinters as adults and harbors the wilt bacterium. Typically in early spring their feeding causes reduced stands, distorted plants, and wilt, especially on small seedlings. CONTROL: Early detection and prompt insecticide sprays.

Spotted and striped cucumber beetle adults and larva *(Diabrotica undecimpunctata howardi and Acalymma vittatum)* on plant with bacterial wilt. The beetles are overwintering sources and vectors of the bacterial wilt pathogen. Wilt often involves only a portion of a plant. CONTROL: Follow general management practices, early detection of beetles and prompt insecticide sprays. Frequent sprays are required.

Squash bug adults, eggs, and nymphs *(Anasa tristis)* and plant damage. Squash bugs are serious pests of squash and may occur on other cucurbits. Adults and nymphs suck plant juices which blackens and dries out vines. CONTROL: Apply insecticide sprays against the immature stages.

Aphid colonies *(Aphis gossypii* and others) and damage. Winged and wingless aphids occur on the underside of leaves. These sucking insects cause little feeding injury, however they spread viruses that cause mosaic. CONTROL: Apply insecticides if populations are very large. This is not effective for controlling mosaic.

Pickleworm *(Diaphania nitidalis)* and characteristic entrance hole with exudate. The movement of pickleworm moths into North Carolina and their presence in specific fields are monitored with sex pheromone traps or occurrence on squash plants. CONTROL: Begin spraying when migration is detected.

Pickleworm *(Diaphania nitidalis)* and relatives:

a. Early instar pickleworm. Note brown head and black spots.

b. Late instar pickleworm. Note brown head and green body.

c. Melonworm *(Diaphania hyalinata)*. Note white stripes on top of body.

d. Pickleworm. Again note brown head with a black notch on the rear of the head capsule and the green body.

e. Squash vine borer *(Melittia cucurbitae)*. Note the black head and the white, deep segmented body.

f. Squash vine borer moth. These moths visit curcurbits during the day.

Fertilizer deficiency. Characteristic symptoms of manganese deficiency.

CONTROL: Apply fertilizers according to annual soil test reports.

Soybean thrips (Sericothrips variabilis) and injury. Feeding winged and wingless, yellow or striped, narrow bodied thrips cause silvery blotches on leaves and reduced plant vigor, especially on small plants during cool, wet periods.

CONTROL: Scout fields for presence of thrips and spray when infestations occur. Use a lOx pocket magnifier and white paper to detect thrips.

Two spotted spider mite adult and egg *(Tetranychus urticae)* and foliar injury Mites suck plant juices causing yellowed leaves and poor plant growth. With population buildup during hot, dry periods, fine webbing may be observed.

CONTROL: Same as for soybean thrips.

Cabbage looper adult, pupa and larva *(Trichoplusia ni)* and injury. Foliar feeding by loopers can strip leaves and delay growth of small plants in late plantings.

CONTROL: Check traps for this migrating insect, scout fields and spray when young larvae are seen.

Deformed fruit often results from poor pollination, extreme weather (too hot, dry, wet, cold) or nutrition levels (too much or little fertilizer).

CONTROL: Place one bee colony per acre after flowering, fertilize according to soil test reports, and irrigate.

Honey bee *(Apis mellifera)*. Honey bees are essential for successful cucumber pollination and production. At least one strong colony per acre is needed to ensure adequate bee activity. Colonies should be placed in the field after the onset of significant flowering.

Gummy Stem Blight is common and is caused by the seed-borne fungus *Mycosphaerella citrullina and* the soil-borne fungus *Phoma exiqua*. All cucurbits are susceptible, and depending on the weather, stem and leafspotting may occur on young plants, especially in late plantings.

CONTROL: Follow general management practices; start spray schedule at first appearance.

Anthracnose is common and is caused by the seedborne fungus *Colletotrichum lagenarium*. Most cucurbits are susceptible. Most commercial cucumber varieties have some resistance. Depending on weather, leaf and fruit spotting may occur on young plants, especially in late plantings.

CONTROL: Follow general management practices; start spray schedule at first appearance.

Angular Leafspot is infrequent and is caused by the seed-borne bacterium *Pseudomonas lachrymans* which can also affect squash and pumpkins. Warm, wet weather may cause stem and leaf spots in midseason.

CONTROL: Follow general management practices; start spray schedule at first appearance; do not enter fields when plants are wet.

Powdery Mildew is caused by two air-borne fungi *Erysiphe cichoracearum and Sphaerotheca fulginea,* but is rarely seen since most commercial cucumbers have excellent resistance. Powdery mildew is often serious in dry weather and is the main cause of early dying of cantaloupes, pumpkins, squash, and greenhouse cucumbers.

CONTROL: Resistant cultivars; sprays; because of fungicide specificity and possible development of resistance in the fungus, switch fungicides if spraying is not effective.

Scab is caused by the seed-borne fungus *Cladosporium cucumerinum* which also affects muskmelons. It is rare because most commercial cucumbers have excellent resistance. Cool night temperatures and high humidities are required. In the active phase, dark, olive-green, velvety mold covers the spot. Leaves, fruit, and young terminals may be affected.

CONTROL: Use resistant varieties, spray at first appearance.

Downy Mildew is caused by the air-borne fungus *Pseudoperonospora cubensis.* Although serious on late season cantaloupes, pumpkins, squash, watermelon, and other cucurbits, it is seldom a problem on cucumbers because most commercial varieties have adequate resistance. The fungus overwinters in Florida, and each year the spores are blown northward and arrive in North Carolina from May to August, if at all. Typically its arrival coincides with summer rains.

CONTROL: Resistant cultivars; start spray schedule at first appearance.

Target Spot is caused by the air-borne fungus Cor*ynespora cassiicola* which attacks many different plants. It is more common on cucumbers in greenhouses than in fields.

CONTROL: Follow general management practices; start spray schedule at first appearance.

Belly Rot is caused by the fungus *Rhizoctonia solani,* present in agricultural soils. It causes damping-off of many crops including cucumbers. Belly rot is common, but its occurrence and intensity are unpredictable.

CONTROL: Avoid problem fields; apply fungicide to soil prior to fruit set; keep harvested cucumbers cool, well ventilated; use chlorine in wash.

Bacterial Wilt is caused by the bacterium *Erwinia tracheiphila.* It overwinters in and is spread by cucumber beetles. Often seen in early to mid-season cucumbers and when temperatures are cool, it occurs on muskmelon and to a lesser extent on squash and pumpkin; watermelons may be immune.

CONTROL: Control the cucumber beetles, spray.

Root-Knot is caused by *Meloidogyne spp.* These nematodes are usually present in agricultural fields and affect many crops. They are especially damaging on plants growing in light, sandy soils when counts are high.

CONTROL: Promptly disk all crops after harvest, treat soil with a fumigant nematicide when warranted by assay report.

Mosaics are caused by viruses. Depicted is Tomato Ringspot, fairly common in the northwestern part of the state. The virus resides in wild brambles and is transmitted by some nematodes. Most commercial cultivars have excellent resistance to Cucumber and Watermelon Mosaic. Until recently mosaics were rarely seen in the eastern counties. Because

Zucchini Yellow Mosaic (closely related to Watermelon Mosaic) may be introduced, it is important to identify all mosaics. Mosaic viruses are transmitted by aphids and related insects.

CONTROL: Select resistant cultivars; avoid problem fields; start spray schedule with stylet oil and insecticides at the seedling stage; use aluminum plastic mulch.

Cottony Leak is caused by the fungus *Pythium spp.* Present in agricultural soils; it causes damping-off of many crops including cucumbers. Cottony leak is common, but its occurrence and intensity are unpredictable.

CONTROL: Avoid problem fields; apply fungicide sprays to soil and foliage prior to fruit set; keep harvested cucumbers cool, well ventilated; use chlorine in wash.

Onion

The onion (*Allium cepa*) (Latin 'cepa' = onion), also known as the bulb onion or common onion, is used as a vegetable and is the most widely cultivated species of the genus *Allium*. This genus also contains several other species variously referred to as onions and cultivated for food, such as the Japanese bunching onion (*A. fistulosum*), the Egyptian onion (*A.* ×*proliferum*), and the Canada onion (*A. canadense*). The name "wild onion" is applied to a number of *Allium* species but *A. cepa* is exclusively known from cultivation and its ancestral wild original form is not known, although escapes from cultivation have become established in some regions. The onion is most frequently a biennial or a perennial plant, but is usually treated as an annual and harvested in its first growing season.

The onion plant has a fan of hollow, bluish-green leaves and the bulb at the base of the plant begins to swell when a certain day-length is reached. In the autumn the foliage dies down and the outer layers of the bulb become dry and brittle.

The crop is harvested and dried and the onions are ready for use or storage. The crop is prone to attack by a number of pests and diseases, particularly the onion fly, the onion eelworm and various fungi that cause rotting. Some varieties of *A. cepa* such as shallots and potato onions produce multiple bulbs.

Onions are cultivated and used around the world. As a foodstuff they are usually served cooked, as a vegetable or part of a prepared savoury dish, but can also be eaten raw or used to make pickles or chutneys. They are pungent when chopped and contain certain chemical substances which irritate the eyes. Onions contain phenolics and flavonoids that have potential anti-inflammatory, anti-cholesterol, anticancer and antioxidant properties.

Description

The onion plant (*Allium cepa*) is unknown in the wild but has been grown and selectively bred in cultivation for at least 7,000 years. It is a biennial plant but is usually grown as an annual. Modern varieties typically grow to a height of 15 to 45 cm (6 to 18 in). The leaves are blueish-green and grow alternately in a flattened, fan-shaped swathe. They are fleshy, hollow and cylindrical, with one flattened side. They are at their broadest about a quarter of the way up beyond which they taper towards a blunt tip. The base of each leaf is a flattened, usually white sheath that grows out of a basal disc. From the underside of the disc, a bundle of fibrous roots extends for a short way into the soil. As the onion matures, food reserves begin to accumulate in the leaf bases and the bulb of the onion swells.

In the autumn the leaves die back and the outer scales of the bulb become dry and brittle, and this is the time at which the crop is normally harvested. If left in the soil over winter, the growing point in the middle of the bulb begins to develop in the spring. New leaves appear and a long, stout, hollow stem expands, topped by a bract protecting a developing inflorescence. The inflorescence takes the form of a globular umbel of white flowers with parts in sixes. The seeds are glossy black and triangular in cross section.

Uses

Historical use: Bulbs from the onion family are thought to have been used as a food source for millennia. In Bronze Age settlements, traces of onion remains were found alongside date stones and fig remains that date back to 5000 BC. However, it is not clear if these were cultivated onions. Archaeological and literary evidence such as the Book of Numbers 11:5 suggests that onions were probably being cultivated around two thousand years later in ancient Egypt, at the same time that leeks and garlic were cultivated. Workers who built the Egyptian pyramids may have been fed radishes and onions.

The onion is easily propagated, transported and stored. The ancient Egyptians worshipped it, believing its spherical shape and concentric rings symbolized eternal life. Onions were even used in Egyptian burials, as evidenced by onion traces being found in the eye sockets of Ramesses IV.

In ancient Greece, athletes ate large quantities of onion because it was believed to lighten the balance of the blood. Roman gladiators were rubbed down with onions to firm up their muscles. In the Middle

Ages, onions were such an important food that people would pay their rent with onions, and even give them as gifts. Doctors were known to prescribe onions to facilitate bowel movements and erections, and to relieve headaches, coughs, snakebite and hair loss.

Onions were taken by the first settlers to North America, where the Native Americans were already using wild onions in a number of ways, eating them raw or cooked in a variety of foods. They also used them to make into syrups, to form poultices and in the preparation of dyes. According to diaries kept by the colonists, bulb onions were one of the first things planted by the Pilgrim Fathers when they cleared the land for cropping in 1648.

Onions were also prescribed by doctors in the early 16th century to help with infertility in women. They were similarly used to raise fertility levels in dogs, cats and cattle, but this was an error as recent research has shown that onions are toxic to dogs, cats, guinea pigs and many other animals.

Culinary Uses

Onions are often chopped and used as an ingredient in various hearty warm dishes, and may also be used as a main ingredient in their own right, for example in French onion soup or onion chutney. They are very versatile and can be baked, boiled, braised, grilled, fried, roasted, sautéed or eaten raw in salads. Their layered nature makes them easy to hollow out once cooked, facilitating stuffing them. Onions are a staple in Indian cuisine, often used as a thickening agent for curries and gravies. Onions pickled in vinegar are eaten as a snack. These are often a side serving in pubs and fish and chip shops throughout the United Kingdom and The Commonwealth, often served with cheese and/or ale in the United Kingdom. In North America, sliced onions are battered and deep fried and served as onion rings.

Onion Types and Products

Common onions are normally available in three colour varieties. *Yellow* or *brown* onions (called *red* in some European countries), are full-flavoured and are the onions of choice for everyday use. Yellow onions turn a rich, dark brown when caramelized and give French onion soup its sweet flavour. The *red* onion (called *purple* in some European countries) is a good choice for fresh use when its colour livens up the dish. It is also used in grilling and char-broiling. *White* onions are the traditional onions that are used in classic Mexican cuisine. They have a golden colour when cooked and a particularly sweet flavour when sautéed.

While the large mature onion bulb is the onion most often eaten, onions can be eaten at immature stages. Young plants may be harvested before bulbing occurs and used whole as spring onions or scallions. When an onion is harvested after bulbing has begun but the onion is not yet mature, the plants are sometimes referred to as *summer* onions.

Additionally, onions may be bred and grown to mature at smaller sizes. Depending on the mature size and the purpose for which the onion is used, these may be referred to as *pearl*, *boiler*, or *pickler* onions, but differ from true pearl onions which are a different species. Pearl and boiler onions may be cooked as a vegetable rather than as an ingredient and pickler onions are often preserved in vinegar as a long-lasting relish.

Onions are available in fresh, frozen, canned, caramelized, pickled and chopped forms. The dehydrated product is available as kibbled, sliced, rings, minced, chopped, granulated and powder forms. Onion powder is a spice widely used when the fresh ingredient is not available. It is made from finely ground, dehydrated onions, mainly the pungent varieties of bulb onions, and has a strong odour. Being dehydrated, it has a long shelf life and comes in several varieties: yellow, red and white.

Non-culinary Uses

Onions have particularly large cells that are readily observed by the human eye at low magnification; consequently, onion tissue is frequently used in science education for demonstrating microscope usage and for learning about the structure of cells.

The pungent juice of onions has been used as a moth repellent and can be rubbed on the skin to prevent insect bites. When applied to the scalp it is said to promote growth of hair and on the face to reduce freckling. It has been used to polish glass and copperware and to prevent rust on iron. If boiling water is poured onto chopped onions and left to cool, the resulting liquor can be sprayed onto plants to increase their resistance to pests, and the onion plants when growing are reputed to keep away moles and insects. Onion skins have been used to produce a yellow-brown dye.

Nutrition and Health

Most onion cultivars are about 89% water, 4% sugar, 1% protein, 2% fibre and 0.1% fat. They contain vitamin C, vitamin B_6, folic acid and numerous other nutrients in small amounts. They are low in fats and in sodium, and with an energy value of 166kJ (40 kcal) per 100 g (3.5 oz) serving, they can contribute their flavour to savoury dishes without raising caloric content appreciably.

Onions contain chemical compounds such as phenolics and flavonoids that basic research shows to have potential anti-inflammatory, anti-cholesterol, anticancer and antioxidant properties. These include quercetin and its glycosides quercetin 3,4'-diglucoside and quercetin-4'-glucoside. There are considerable differences between different varieties in potential antioxidant content. Shallots have the highest level, six times the amount found in Vidalia onions, the variety with the smallest amount.

Some people suffer from allergic reactions after handling onions. Symptoms can include contact dermatitis, intense itching, rhinoconjunctivitis, blurred vision, bronchial asthma, sweating and anaphylaxis. There may be no allergic reaction in these individuals to the consumption of onions, perhaps because of the denaturing of the proteins involved during the cooking process.

While onions and other members of the genus *Allium* are commonly consumed by humans, they can be deadly for dogs, cats, guinea pigs, monkeys and other animals. The toxicity is caused by the sulfoxides present in raw and cooked onions which many animals are unable to digest. Ingestion results in anaemia caused by the distortion and rupture of red blood cells. Sick pets are sometimes fed with tinned baby foods and any that contain onion should be avoided. Nor is it good for pets to be fed onion-containing leftovers such as pizza, canned spaghetti, Chinese dishes and onion rings. The typical toxic doses are 5 g (0.2 oz) per kg (2.2 lb) bodyweight for cats and 15 to 30 g (0.5 to 1.1 oz) per kg for dogs.

In India, some sects do not eat onions as they believe them to be an aphrodisiac. Various schools of Buddhism also advise against the consumption of onions and garlic because they believe that they increase desire when eaten cooked and anger when eaten raw.

Cultivation

Onions are best cultivated in fertile soils that are well-drained. Sandy loams are good as they are low in sulphur, while clayey soils usually have a high sulphur content and produce pungent bulbs. Onions require a high level of nutrients in the soil. Phosphorus is often present in sufficient quantities but may be applied before planting because of its low level of availability in cold soils. Nitrogen and potash can be applied at intervals during the growing season, the last application of nitrogen being at least four weeks before harvesting. Bulbing onions are day-length sensitive; their bulbs begin growing only after the number of daylight hours has surpassed some minimal quantity. Most traditional

European onions are what is referred to as "long-day" onions, producing bulbs only after 14+ hours of daylight occurs. Southern European and North African varieties are often known as "intermediate day" types, requiring only 12–13 hours of daylight to stimulate bulb formation. Finally, "short-day" onions, which have been developed in more recent times, are planted in mild-winter areas in the fall and form bulbs in the early spring, and require only 11–12 hours of daylight to stimulate bulb formation. Onions are a cool-weather crop and can be grown in USDA zones 3 to 9. Hot temperatures or other stressful conditions cause them to "bolt", meaning that a flower stem begins to grow.

Onions may be grown from seed or from sets. Onion seeds are short lived and fresh seed germinates better. The seeds are sewn thinly in shallow drills, thinning the plants in stages. In suitable climates, certain cultivars can be sown in late summer and autumn to overwinter in the ground and produce early crops the following year. Onion sets are produced by sowing seed thickly in early summer in poor soil and the small bulbs produced are harvested in the autumn. These bulbs are planted the following spring and grow into mature bulbs later in the year. Certain cultivars are used for this purpose and these may not have such good storage characteristics as those grown directly from seed.

Routine care during the growing season involves keeping the rows free of competing weeds, especially when the plants are young. The plants are shallow rooted and do not need a great deal of water when established. Bulbing usually takes place after twelve to eighteen weeks. The bulbs can be gathered when needed to eat fresh but if required to be kept in store, should be harvested after the leaves have died back naturally. In dry weather they can be left on the surface of the soil for a few days to dry out properly, then they can be placed in nets, roped into strings or laid in layers in shallow boxes. They should be stored in a well-ventilated, cool place such as a shed.

Pests and Diseases

Onions suffer from a number of plant disorders. The most serious for the home gardener are likely to be the onion fly, stem and bulb eelworm, white rot and neck rot. Diseases affecting the foliage include rust and smut, downy mildew and white tip disease. The bulbs may be affected by splitting, white rot and neck rot. Shanking is a condition when the central leaves turn yellow and the inner part of the bulb collapses into an unpleasant-smelling slime. Most of these disorders are best treated by removing and burning affected plants. The larvae of the onion leaf miner or leek moth (*Acrolepiopsis assectella*) sometimes attacks the foliage and may burrow down into the bulb.

The onion fly (*Delia antiqua*) lays eggs on the leaves and stems and on the ground close to onion, shallot, leek and garlic plants. The fly is attracted to the crop by the smell of damaged tissue and is liable to occur after thinning. Plants grown from sets are less prone to attack.The larvae tunnel into the bulbs and the foliage wilts and turns yellow. The bulbs are disfigured and rot, especially in wet weather. Control measures may include crop rotation, the use of seed dressings, early sowing or planting and the removal of infested plants.

The onion eelworm (*Ditylenchus dipsaci*), a tiny parasitic soil-living nematode, causes swollen distorted foliage. Young plants are killed and older ones produce soft bulbs. There is no cure and affected plants should be uprooted and burnt. The site should not be used for growing onions again for several years and should also be avoided for growing carrots, parsnips and beans which are also susceptible to the eelworm.

White rot of onions, leeks and garlic is caused by the soil-borne fungus *Sclerotium cepivorum*. As the roots rot, the foliage turns yellow and wilts. The base of the bulbs is attacked and becomes covered by a fluffy white mass of mycelia which later produces small, globular black structures called sclerotia. These resting structures remain in the soil to reinfect a future crop. There is no cure for this fungal disease so affected plants should be removed and destroyed and the ground used for unrelated crops in subsequent years.

Neck rot is a fungal disease affecting onions in storage. It is caused by *Botrytis allii* which attacks the neck and upper parts of the bulb causing a grey mould to develop. The symptoms often first occur where the bulb has been damaged and spread downwards in the affected scales. Large quantities of spores are produced and crust-like sclerotia may also develop. In time a dry rot sets in and the bulb becomes a dry, mummified structure. This disease may be present throughout the growing period but only manifest itself when the bulb is in store. Anti-fungal seed dressings are available and the disease can be minimised by preventing physical damage to the bulbs at harvesting, careful drying and curing of the mature onions and correct storage in a cool, dry place with plenty of circulating air.

Storage in the Home

Cooking onions and sweet onions are better stored at room temperature, optimally in a single layer, in mesh bag in a dry, cool, dark, well-ventilated location. In this environment, cooking onions have a shelf life of three to four weeks and sweet onions one to two weeks. Cooking onions will absorb odours from apples and pears. Also, they draw moisture from vegetables with which they are stored which may cause them to decay.

Sweet onions have a greater water and sugar content than cooking onions. This makes them sweeter and milder tasting but reduces their shelf life. Sweet onions can be stored refrigerated; they have a shelf life of approximately one month. Irrespective of type, any cut pieces of onion are best tightly wrapped, stored away from other produce, and used within two to three days.

Varieties

Common Onion Group (var. cepa): Most of the diversity within *A. cepa* occurs within this group, the most economically important *Allium* crop. Plants within this group form large single bulbs, and are grown from seed or seed-grown sets. The majority of cultivars grown for dry bulbs, salad onions, and pickling onions belong to this group. The range of diversity found among these cultivars includes variation in photoperiod (length of day that triggers bulbing), storage life, flavour, and skin colour. Common onions range from the pungent varieties used for dried soups and onion powder to the mild and hearty sweet onions, such as the Vidalia from Georgia, USA, or Walla Walla from Washington that can be sliced and eaten on a sandwich instead of meat.

Aggregatum Group (var. aggregatum): This group contains shallots and potato onions, also referred to as multiplier onions. The bulbs are smaller than those of common onions, and a single plant forms an aggregate cluster of several bulbs. They are propagated almost exclusively from daughter bulbs, although reproduction from seed is possible. Shallots are the most important subgroup within this group and comprise the only cultivars cultivated commercially. They form aggregate clusters of small, narrowly ovoid to pear-shaped bulbs. Potato onions differ from shallots in forming larger bulbs with fewer bulbs per cluster, and having a flattened (onion-like) shape. However, intermediate forms exist.

I'itoi onion (*Allium cepa*) is a prolific multiplier onion cultivated in the Baboquivari Peak Wilderness, Arizona area. This small bulb type has a shallot-like flavour and is easy to grow and ideal for hot, dry climates. Bulbs are separated, and planted in the fall 1 inch below surface and 12 inches apart. Bulbs will multiply into clumps and can be harvested throughout the cooler months. Tops will die back in the heat of summer and may return with heavy rains; bulbs can remain in the ground or be harvested and stored in a cool dry place for planting in the fall. The plants rarely flower; propagation is by division.

Hybrids with A. cepa Parentage: A number of hybrids are cultivated that have *A. cepa* parentage, such as the diploid tree onion or Egyptian onion (*A.* ×*proliferum*), Wakegi onion (*A.* ×*wakegi*), and the triploid onion (*A.* ×*cornutum*).

The tree onion or Egyptian onion produces bulblets in the umbel instead of flowers, and is now known to be a hybrid of *A. cepa* × *A. fistulosum*. It has previously been treated as a variety of *A. cepa*, for example *A. cepa* var. *proliferum*, *A. cepa* var. *bulbiferum*, and *A. cepa* var. *viviparum*.

The Wakegi onion is also known to be a hybrid between *A. cepa* and *A. fistulosum*, with the *A. cepa* parent believed to be from the Aggregatum Group of cultivars. It has been grown for centuries in Japan and China for use as a salad onion.

Under the rules of botanical nomenclature, both the Egyptian onion and Wakegi onion should be combined into one hybrid species, having the same parent species. Where this is followed, the Egyptian onion is named *A.* ×*proliferum* Eurasian Group and the Wakegi onion is named *A.* ×*proliferum* East Asian Group.

The triploid onion is a hybrid species with three sets of chromosomes, two sets from *A. cepa* and the third set from an unknown parent. Various clones of the triploid onion are grown locally in different regions, such as 'Ljutika' in Croatia, and 'Pran', 'Poonch' and 'Srinagar' in the India-Kashmir region. 'Pran' is grown extensively in the Northern Indian provinces of Jammu and Kashmir. There are very small genetic differences between 'Pran' and the Croatian clone 'Ljutika', implying a monophyletic origin for this species.

Some authors have used the name *A. cepa* var. *viviparum* (Metzg.) Alef. for the triploid onion, but this name has also been applied to the Egyptian onion. The only name unambiguously connected with the triploid onion is *A.* ×*cornutum*.

Scallions or salad onions may be grown from the Welsh onion (*A. fistulosum*) as well as from *A. cepa*. Young plants of *A. fistulosum* and *A. cepa* look very similar, but may be distinguished by their leaves, which are circular in cross-section in *A. fistulosum* rather than flattened on one side.

Production and Trade

It is estimated that around the World, over 9,000,000 acres (3,642,000 ha) of onions are grown annually. About 170 countries cultivate onions for domestic use and about eight percent of the global production is traded internationally.

Garlic

A world of flavours awaits in every bulb! Garlic's taste has several dimensions that come alive depending on how the plant is cooked.

Garlic (Allium sativum L.) is in the onion family. For thousands of years it has been grown for culinary and medicinal uses. Garlic grows in a wide range of conditions; while most production is in mild areas, like California, some varieties grow well in cold climates, often with better flavour.

Types and Varieties

There are two types of garlic, hardneck (or topset) and softneck. Hardneck garlic produces false flower stalks, called scapes. It typically has a half dozen cloves per bulb, plus or minus, while softneck types have twice that many. Softneck garlic dominates commodity production but in New England many growers prefer hardneck types for retail sale due to their flavour and appearance, which often includes a reddish purple clove covering versus the white colour common to softneck types. Softneck garlic generally has a longer storage life than hardneck and is easier to braid. After centuries of cultivation garlic has lost the ability to produce seeds so it is vegetatively propagated by saving bulbs from year to year and planting the cloves. The small bulbils produced on hardneck scapes can be used for propagation, but it takes several years of planting and selection to achieve marketable size bulbs.

Although many different variety names are used in garlic commerce, recent genetic research suggests there are only about 10 major varieties of garlic. These express different characteristics from one location to another, complicating variety identification. Since there is no standardization of varieties, as with potatoes, one must take garlic variety names with a grain of salt. It's a good idea to start out with several different varieties produced in your area select and save those that perform best.

Soil Fertility

A well-drained soil with good tilth and plenty of organic matter is ideal for garlic. Excess moisture, compaction or droughty conditions will reduce yields. The optimum soil pH is between 6 and 7. Since garlic commences growth very early in the season it is important to avoid field that are slow to drain in the spring.

Planting

Garlic is planted in the fall since it requires a cold treatment to induce bulbs formation. Planting typically occurs from October in northern New England to early November in southern areas. The goal is to allow good development of roots but not enough time for the shoots to emerge from soil before winter. Many different planting arrangements are used by growers depending on their irrigation, mulching and weed control systems; spacing too close can reduce bulb size while spacing too far apart reduces yield per area of land. Planting arrangements include: 2 row beds 30 inches apart on centre with 6 inch spacing in and between rows; 3 or 4 rows per bed, with 6 to 8 inches between and within rows; single rows spaced 24 to 30 inches with 6 inch spacing in the row. Wide row spacing between rows allows for easy mechanical cultivation for weed control; multiple rows per bed allow for use of plastic much to control weeds.

Garlic varieties differ in size and weight of cloves; generally, there are about 50 cloves per pound of cloves. Large cloves tend to produce the most vigorous plants and largest bulbs so small cloves are often not planted. Bulbs should be separated no more than a day or two before planting so they do not dry out. Cloves should always be planted with the root side down, about three inches deep. Only healthy-looking cloves should be planted to avoid disease and nematode problems.

Field Culture

A layer of clean straw mulch is typically applied to garlic to avoid soil temperature fluctuations in the winter and early spring. The mulch also will help conserve soil moisture and suppress weeds. However, it may be advisable to remove mulch in a very wet spring to allow soil to dry out and thus reduce the threat of soil borne diseases. In very cold growing areas, removal of mulch can also speed soil warming and garlic growth in the spring. Because garlic is shallow rooted irrigation is very beneficial to maintain a consistent water supply, and yields will suffer if it is not applied during dry periods. Removal of scapes from hardneck garlic once they fully emerge may improve bulb yield, especially on soils of low fertility; the scapes are edible.

Harvest and Storage

Although variable depending on variety and growing conditions, one pound of garlic 'seed ' bulbs will usually yield between four and eight pounds at harvest. Depending on the varieties grown they may be ready to harvest over several weeks from late June to mid-July. When the lower third of leaves turn brown pull some bulbs, cut them across the cloves and look to see if the cloves have filled the skins and are ready to harvest. Pull, dig and/or undercut the bulbs to remove them. Unless a lot of soil is adhered to the bulbs they do not need to be washed at harvest although your markets may demand it. Place the harvested plants on wire racks or tie up in bundles and hang to cure for several weeks in a dry area with good ventilation. Then cut the tops off leaving an inch or so and trim the roots closely. If necessary bulbs may be brushed or the outer skin gently rubbed off to clean them.

Seed garlic should be stored at 50 degrees with relative humidity of 65-70%. Cloves sprout most rapidly between 40 to 50 degrees. Garlic for table stock should be stored at 32 degrees and 65-70% relative humidity. Well cured bulbs in good condition should keep for 6 to 7 months. Relative humidity in storage is lower than for most vegetables because high humidity causes root growth and mold.

Beans

Bean is a common name for large plant seeds used for human food or animal feed of several genera of the family Fabaceae (alternately Leguminosae). The term *bean* originally referred to the seed of the broad or fava bean, but was later expanded to include members of the New World genus *Phaseolus*, such as the common bean and the runner bean, and the related genus *Vigna*. The term is now applied generally to many other related plants such as Old World soybeans, peas, chickpeas (garbanzos), vetches, and lupins.

Bean is sometimes used as a synonym of *pulse*, an edible legume, though the term *pulses* is normally reserved for leguminous crops harvested for their dry grain. The term *bean* usually excludes crops used mainly for oil extraction (such as soybeans and peanuts), as well as those used exclusively for sowing purposes (such as clover and alfalfa). Leguminous crops harvested green for food, such as snap peas, snow peas, and so on, are not considered beans, and are classified as vegetable crops. According to United Nation's Food and Agriculture Organization the term *bean* should include only species of *Phaseolus*; however, a strict consensus definition has proven difficult because in the past, several species such as *Vigna angularis* (azuki bean), *mungo* (black gram), *radiata* (green gram), *aconitifolia* (moth bean)) were classified as *Phaseolus* and later reclassified. The use of the term *bean* to refer to species other than *Phaseolus* thus remains. In some countries, the term *bean* can mean a host of different species.

In English usage, the word *bean* is also sometimes used to refer to the seeds or pods of plants that are not in the family leguminosae, but which bear a superficial resemblance to true beans—for example coffee beans, castor beans and cocoa beans (which resemble bean seeds), and vanilla beans, which superficially resemble bean pods.

Cultivation

Unlike the closely related pea, beans are a summer crop that needs warm temperatures to grow. Maturity is typically 55–60 days from planting to harvest. As the bean pods mature, they turn yellow and dry up, and the beans inside change from green to their mature colour. As a vine, bean plants need external support, which may be provided in the form of special "bean cages" or poles. Native Americans customarily grew them along with corn and squash (the so-called Three Sisters), with the tall cornstalks acting as support for the beans.

In more recent times, the so-called "bush bean" has been developed which does not require support and has all its pods develop simultaneously (as opposed to pole beans which develop gradually). This makes the bush bean more practical for commercial production.

History

Beans are one of the longest-cultivated plants. Broad beans, also called fava beans, in their wild state the size of a small fingernail, were gathered in Afghanistan and the Himalayan foothills. In a form improved from naturally occurring types, they were grown in Thailand since the early seventh millennium BCE, predating ceramics. They were deposited with the dead in ancient Egypt. Not until the second millennium BCE did cultivated, large-seeded broad beans appear in the Aegean, Iberia and transalpine Europe. In the *Iliad* (late-8th century) is a passing mention of beans and chickpeas cast on the threshing floor.

Beans were an important source of protein throughout Old and New World history, and still are today.

The oldest-known domesticated beans in the Americas were found in Guitarrero Cave, an archaeological site in Peru, and dated to around the second millennium BCE.

Most of the kinds commonly eaten fresh or dried, those of the genus *Phaseolus*, come originally from the Americas, being first seen by a European when Christopher Columbus, during his exploration of what may have been the Bahamas, found them growing in fields. Five kinds of *Phaseolus* beans were domesticated by pre-Columbian peoples: common beans (*Phaseolus vulgaris*) grown from Chile to the northern part of what is now the United States, and lima and sieva beans (*Phaseolus lunatus*), as well as the less widely distributed teparies (*Phaseolus acutifolius*), scarlet runner beans (*Phaseolus coccineus*) and polyanthus beans (*Phaseolus polyanthus*) One especially famous use of beans by pre-Columbian people as far north as the Atlantic seaboard is the "Three Sisters" method of companion plant cultivation:

In the New World, many tribes would grow beans together with maize (corn), and squash. The corn would not be planted in rows as is done by European agriculture, but in a checkerboard/hex fashion across a field, in separate patches of one to six stalks each.

Beans would be planted around the base of the developing stalks, and would vine their way up as the stalks grew. All American beans at that time were vine plants, "bush beans" having been bred only more recently. The cornstalks would work as a trellis for the beans, and the beans would provide much-needed nitrogen for the corn.

Squash would be planted in the spaces between the patches of corn in the field. They would be provided slight shelter from the sun by the corn, would shade the soil and reduce evaporation, and would deter many animals from attacking the corn and beans because their coarse, hairy vines and broad, stiff leaves are difficult or uncomfortable for animals such as deer and raccoons to walk through, crows to land on, etc.

Dry beans come from both Old World varieties of broad beans (fava beans) and New World varieties (kidney, black, cranberry, pinto, navy/haricot).

Beans are a heliotropic plant, meaning that the leaves tilt throughout the day to face the sun. At night, they go into a folded "sleep" position.

Types

Currently, the world genebanks hold about 40,000 bean varieties, although only a fraction are mass-produced for regular consumption.

Some bean types include:

- *Vicia*
  - *V. faba* or broad bean

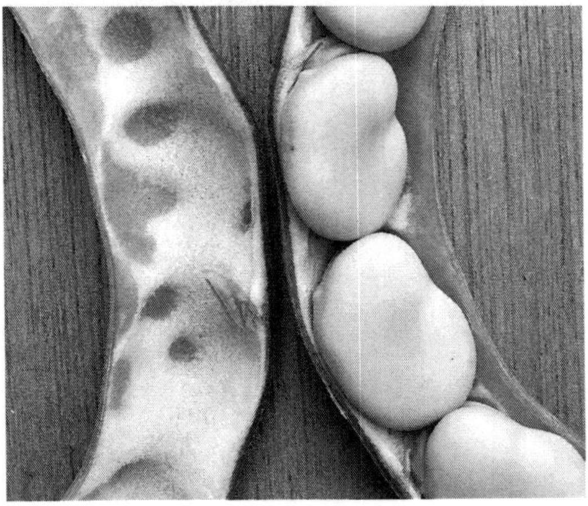

Figure: *Vica faba or broad beans, known in the US as fava beans*

- *Vigna*
  - o *V. aconitifolia* or Moth bean
  - o *V. angularis* or azuki bean
  - o *V. mungo* or urad bean
  - o *V. radiata* or mung bean
  - o *V. subterranea*, Bambara bean or ground-bean
  - o *V. umbellatta* or ricebean
  - o *V. unguiculata* or cowpea (includes the black-eyed pea, yardlong bean and others)
- *Cicer*
  - o *C. arietinum* or chickpea (also known as the garbanzo bean)
- *Pisum*
  - o *P. sativum* or pea
- *Lathyrus*
  - o *Lathyrus sativus* (Indian pea)
  - o *Lathyrus tuberosus* (Tuberous pea)
- *Lens*

Figure: *Lentils*

  - o *L. culinaris* or lentil
- *Lablab*
  - o *L. purpureus* or hyacinth bean
- *Phaseolus*
  - o *P. acutifolius* or tepary bean
  - o *P. coccineus* or runner bean
  - o *P. lunatus* or lima bean

  - *P. vulgaris* or common bean (includes the pinto bean, kidney bean, black bean, as well as green beans, and many others)
  - *P. polyanthus* (aka P. Dumosus, recognized as a separate species in 1995)
- *Glycine*
  - *G. max* or soybean
- *Psophocarpus*
  - *P. tetragonolobus* or winged bean
- *Cajanus*
  - *C. cajan* or pigeon pea
- *Stizolobium*
  - *S.* spp or velvet bean
- *Cyamopsis*
  - *C. tetragonoloba* or guar
- *Canavalia*
  - *C. ensiformis* or jack bean
  - *Canavalia gladiata* or sword bean
- *Macrotyloma*
  - *M. uniflorum* or horse gram
- *Lupinus* or Lupin
  - *L. mutabilis* or tarwi
  - *Lupinus albus* or lupini bean
- *Erythrina*
  - *E. herbacea* or Coral bean
- *Arachis*
  - *A. hypogaea* or Peanut

Toxins

Some kinds of raw beans, especially red and kidney beans, contain a harmful toxin (lectin phytohaemagglutinin) that must be removed by cooking. A recommended method is to boil the beans for at least ten minutes; undercooked beans may be more toxic than raw beans. Cooking beans in a slow cooker, because of the lower temperatures often used, may not destroy toxins even though the beans do not smell or taste 'bad' (though this should not be a problem if the food reaches boiling temperature and stays there for some time).

Fermentation is used in some parts of Africa to improve the nutritional value of beans by removing toxins. Inexpensive fermentation improves the nutritional impact of flour from dry beans and improves digestibility, according to research co-authored by Emire Shimelis, from the Food Engineering Programme at Addis Ababa University. Beans are a major source of dietary protein in Kenya, Malawi, Tanzania, Uganda and Zambia.

Nutrition

Beans have significant amounts of fibre and soluble fibre, with one cup of cooked beans providing between nine and 13 grams of fibre. Soluble fibre can help lower blood cholesterol. Beans are also high in protein, complex carbohydrates, folate, and iron.

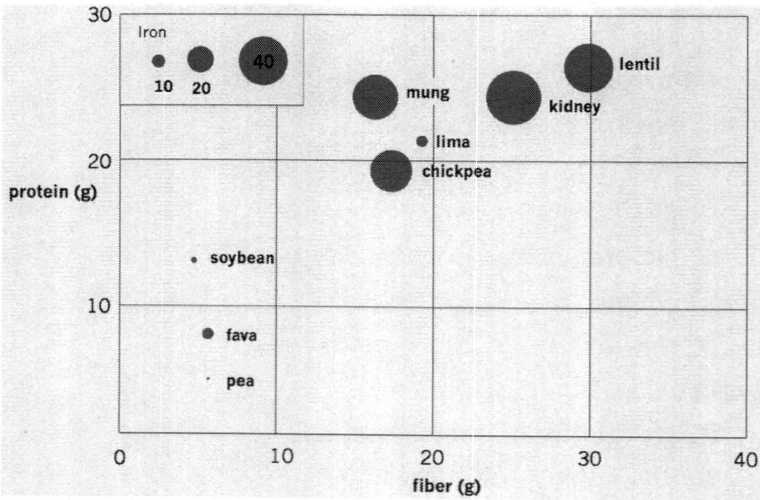

Figure: *This figure shows the grams of fibre and protein per 100g serving of each legume. The size of the circle is proportional to its iron content. From this view, lentil and kidney beans are by far the healthiest while soybeans and peas have the least nutrients per serving.*

Flatulence

Many edible beans, including broad beans and soybeans, contain oligosaccharides (particularly raffinose and stachyose), a type of sugar molecule also found in cabbage. An anti-oligosaccharide enzyme is necessary to properly digest these sugar molecules. As a normal human digestive tract does not contain any anti-oligosaccharide enzymes, consumed oligosaccharides are typically digested by bacteria in the large intestine. This digestion process produces flatulence-causing gases as a byproduct. This aspect of bean digestion is the basis for the children's rhyme "Beans, Beans, the Musical Fruit".

Some species of mold produce alpha-galactosidase, an anti-oligosaccharide enzyme, which humans can take to facilitate digestion of oligosaccharides in the small intestine. This enzyme, currently sold in the United States under the brand-names Beano and Gas-X Prevention, can be added to food or consumed separately. In many cuisines beans are cooked along with natural carminatives such as anise seeds, coriander seeds and cumin.

One effective strategy is to soak beans in alkaline (baking soda) water overnight before rinsing thoroughly. Sometimes vinegar is added, but only after the beans are cooked as vinegar interferes with the beans' softening. Fermented beans will usually not produce most of the intestinal problems that unfermented beans will, since yeast can consume the offending sugars.

Production

***Figure:** Lablab bean and bean flower cultivated in West Bengal, India*

The world leader in production of dry bean is India, followed by Brazil and Myanmar. In Africa, the most important producer is Tanzania.

Common Beans

Phaseolus vulgaris, the common bean, string bean, field bean, flageolet bean, French bean, garden bean, haricot bean, pop bean, or snap bean, is a herbaceous annual plant grown worldwide for its edible fruit, either the dry seed or the unripe fruit, both of which are referred to as beans. The leaf is also occasionally used as a vegetable, and the straw can be used for fodder. Along with other species of the bean genus (*Phaseolus*), it is classified botanically into the legume family (Fabaceae), most of whose members acquire nitrogen through an association with rhizobia, a species of nitrogen-fixing bacteria.

The common bean is a highly variable species with a long history of cultivation. All of the wild members of the species have a climbing habit, but there are many cultivars, classified as bush beans, or pole beans depending on their style of growth. These include the kidney bean, the navy bean, the pinto bean, and the wax bean. The other major types of commercially-grown bean are the runner bean (*Phaseolus coccineus*) and the broad bean (*Vicia faba*).

Production of beans is well distributed worldwide, with countries in Asia, Africa, Europe, Oceania, South and North America among the top bean growers. Brazil and India are the largest producers of dry beans, while China produces, by far, the largest quantity of green beans. Worldwide, 23 million tonnes of dry common beans and 17.1 million tonnes of green beans were grown in 2010.

The wild *Phaseolus vulgaris* was native to the Americas and was domesticated separately in Mesoamerica and in the southern Andes region, giving the domesticated bean two gene pools which remain separate to this day. Along with squash and maize, beans are one of the "Three Sisters" central to indigenous North American agriculture.

Description

The common bean is a highly variable species with a long history. Bush varieties form erect bushes 20–60 cm (8–20 in) tall, while pole or running varieties form vines 2–3 m (7–10 ft) long. All varieties bear alternate, green or purple leaves, which are divided into three oval, smooth-edged leaflets, each 6–15 cm (2–6 in) long and 3–11 cm (1–4 in) wide. The white, pink, or purple flowers are about 1 cm long, and they give way to pods 8–20 cm (3–8 in) long and 1–1.5 cm wide. These may be green, yellow, black, or purple in colour, each containing 4–6 beans. The beans are smooth, plump, kidney-shaped, up to 1.5 cm long, range widely in colour, and are often mottled in two or more colours.

Dry Beans

Similar to other beans, the common bean is high in starch, protein and dietary fibre and is an excellent source of iron, potassium, selenium, molybdenum, thiamine, vitamin B6, and folate.

Dry beans will keep indefinitely if stored in a cool, dry place, but as time passes, their nutritive value and flavour degrade and cooking times lengthen. Dried beans are almost always cooked by boiling, often after being soaked in water for several hours. While the soaking is not strictly necessary, it shortens cooking time and results in more evenly textured beans. In addition, soaking beans removes 5 to 10 percent of the gas-producing sugars that can cause flatulence for some people. The methods include simple overnight soaking and the power soak method in which beans are boiled for three minutes and then set aside for 2–4 hours.

Before cooking, the soaking water is drained off and discarded. Dry common beans take longer to cook than most pulses: cooking times vary from one to four hours, but are substantially reduced with pressure cooking.

In Mexico, Central America and South America the traditional spice used with beans is *epazote*, which is also said to aid digestion. In East Asia a type of seaweed, *kombu*, is added to beans as they cook for the same purpose. Salt, sugar, and acidic foods such as tomatoes may harden uncooked beans, resulting in seasoned beans at the expense of slightly longer cooking times.

Dry beans may also be bought cooked and canned as refried beans, or whole with water, salt, and sometimes sugar.

Green Beans and Wax Beans

Figure: *Bean cultivars including three wax beans*

Green beans, also known as string bean, snap bean in the northeastern and western United States, or ejotes in Mexico, are the unripe fruit of various cultivars of the common bean (*Phaseolus vulgaris*). Green bean cultivars have been selected especially for the fleshiness, flavour, or sweetness of their pods.

Haricots verts, French for "green beans", also known as French beans, French green beans, French filet beans, Fine beans (British

English), is a variety of green bean that is longer, thinner, crisp, and tender. It is different from the haricot bean, which is a dry bean.

The three commonly known types of green beans are: string or snap beans, which may be round or have a flat pod; stringless or French beans, which lack a tough, fibrous "string" running along the length of the pod; and runner beans, which belong to a separate species, *Phaseolus coccineus*. Green beans may have a purple rather than green pod, which changes to green when cooked. Wax beans are *Phaseolus vulgaris* beans that have a yellow or white pod. Wax bean cultivars are commonly grown; the plants are often of the bush form.

Compared to dry beans, green and wax beans provide less starch and protein and more vitamin A and vitamin C. Green beans and wax beans are often steamed, boiled, stir-fried, or baked in casseroles.

Culinary Use

Green beans are of nearly universal distribution. They are marketed canned, frozen, and fresh. Green beans are often steamed, boiled, stir-fried, or baked in casseroles. A dish with green beans popular throughout the United States, particularly at Thanksgiving, is green bean casserole, which consists of green beans, cream of mushroom soup, and French fried onions.

Some restaurants in the USA serve green beans that are battered and fried, and Japanese restaurants in the United States frequently serve green bean tempura. Green beans are also sold dried and fried with vegetables such as carrots, corn, and peas.

Beans contain high concentrations of lectins and may be harmful if consumed in excess in uncooked or improperly cooked form.

The flavonol miquelianin (Quercetin 3-O-glucuronide) can be found in green beans.

Cultivation

Green beans are found in two major groups, bush beans and pole beans.

Bush beans are short plants, growing to approximately two feet in height, without requiring supports. They generally reach maturity and produce all of their fruit in a relatively short period of time, then cease to produce. Gardeners may grow more than one crop of bush beans in a season.

Pole beans have a climbing habit and produce a twisting vine. Runner beans have a similar habit but are a different species of bean.

Varieties

Over 130 varieties of green bean are known. Varieties specialized for use as green beans, selected for the succulence and flavour of their pods, are the ones usually grown in the home vegetable garden, and many varieties exist. Pod colour can be green, purple, red, or streaked. Shapes range from thin "fillet" types to wide "romano" types and more common types in between.

The following varieties are among the most common and widely grown in the USA. Closely related varieties are listed on the same line.

Bush Types

- Bountiful, 50 days (green, heirloom)
- Burpee's Stringless Green Pod, 50 days (green, heirloom)
- Contender, 50 days (green)
- Topcrop, 51 days (green), 1950 AAS winner
- Red Swan, 55 days (red)
- Blue Lake 274, 58 days (green)
- Maxibel, 59 days (green fillet)
- Roma II, 59 days (green romano)
- Improved Commodore / Bush Kentucky Wonder, 60 days (green), 1945 AAS winner
- Dragon's Tongue, 60 days (streaked)
- Jade / Jade II, 65 days (green)

Pole Types

- Blue Lake, 60 days (green)
- Fortex, 60 days (green fillet)
- Kentucky Blue, 63 days (green), 1991 AAS winner
- Old Homestead / Kentucky Wonder, 65 days (green, heirloom)
- Rattlesnake, 72 days (streaked, heirloom)
- Purple King, 75 days (purple)

Shelling Beans

Shell, shelled, or shelling beans are beans removed from their pods before being cooked or dried. Common beans can be used as shell beans, but the term also refers to other species of beans whose pods are not typically eaten, such as lima beans, soybeans, peas, and fava beans. Fresh shell beans are nutritionally similar to dry beans, but are prepared more like a vegetable, often being steamed, fried, or made into soups.

Popping Beans

The nuña is an Andean subspecies, *Phaseolus vulgaris* subsp. *nunas* (formerly *Phaseolus vulgaris* Nuñas Group), with round, multicoloured seeds that resemble pigeon eggs. When cooked on high heat, the bean explodes, exposing the inner part, in the manner of popcorn and other puffed grains.

Toxicity

The toxic compound phytohaemagglutinin, a lectin, is present in many common bean varieties, but is especially concentrated in red kidney beans. White kidney beans contain about a third as much toxin as the red variety; broad beans (*Vicia faba*) contain 5 to 10% as much as red kidney beans.

Phytohaemagglutinin can be deactivated by boiling beans; ten minutes at boiling point (100 °C (212 °F)) are sufficient to degrade the toxin, but not to cook the beans, the U.S Food and Drug Administration recommends boiling for 30 minutes to ensure they reach a sufficient temperature for long enough to completely destroy the toxin. For dry beans, the FDA also recommends an initial soak of at least 5 hours in water which should then be discarded. Outbreaks of poisoning have been associated with cooking kidney beans in slow cookers.

The primary symptoms of phytohaemagglutinin poisoning are nausea, vomiting, and diarrhea. Onset is from one to three hours after consumption of improperly prepared beans, and symptoms typically resolve within a few hours. Consumption of as few as four or five raw, soaked kidney beans can cause symptoms.

Beans are high in purines, which are metabolized to uric acid. Uric acid is not a toxin as such, but may promote the development or exacerbation of gout. For this reason, people with gout have been advised in the past to limit their consumption of beans. However, more recent research has questioned this association, finding that moderate intake of purine-rich foods is not associated with increased risk of gout.

Canned red kidney beans, on the other hand, are safe to use immediately.

Varieties

Many well-known bean varieties belong to this species, and none of the lists below are in any way exhaustive. Both bush and running (pole) varieties exist. The colours and shapes of pods and seeds vary over a wide range.

Black Turtle Beans

The Black turtle bean has small, shiny black seeds. It is especially popular in Latin American cuisine.

Cranberry and Borlotti Beans

Cranberry beans originated in Colombia as the cargamanto bean. Borlotti or Roman beans are a variety of cranberry bean bred in Italy to have a thicker skin. They are much used in Mediterranean cuisine.

Flageolet Beans

Flageolet beans are often eaten in France. The seeds are small, light green, and kidney shaped. The texture is firm yet creamy if shelled and cooked when fresh but semi-dry.

Kidney Beans

Kidney beans, also known as red beans, are named for their visual resemblance in shape and colour to kidneys. They are commonly used in chili soup chili con carne and are an integral part of the cuisine in northern regions of India. They are also used in New Orleans and much of southern Louisiana for the classic Monday Creole dish of red beans and rice.

Pea Beans

A type of *Phaseolus vulgaris* called pea bean has been recorded in Britain since the 16th century. In the US the name "pea bean" is also used to describe small white beans, and the same name is used for *Vigna unguiculata* subsp. *sesquipedalis*, also called yard-long bean and cowpea. The seeds of the British pea bean are bicoloured red-brown and white. The plants are a typical climbing bean. The beans are either eaten in the pod like French beans or they may be harvested when mature and eaten as other dried beans.

Pink Beans

Pink beans are small, pale pink, oval-shaped beans also known by the Spanish name *habichuelas rosadas*. The Santa Maria *pinquito* (Spanglish = pink and small), is commercially grown on the mesas above Santa Maria, California, and is a necessary ingredient in Santa Maria Style BBQ.

Pinto Beans

Pinto beans are named for their mottled skin. They are the most common bean in the United States and northwestern Mexico, and are most often eaten whole in broth or mashed and refried. Either whole or mashed, they are a common filling for burritos. The young pods may also be harvested and cooked as green pinto beans.

White Beans

Navy beans or haricot beans are particularly popular in the United Kingdom and the United States. Other white beans include cannellini, a popular variety in central and southern Italy that is related to the kidney bean. White beans are the most abundant plant-based source of phosphatidylserine known.

Yellow Beans

Sinaloa Azufrado, Sulphur, Mayocoba, and Peruano (also called canary) are types of yellow beans.

Peruano beans (also called canary or mayacoba beans) are small, oval, yellow beans about 1/2 inch (1 cm) long with a thin skin. They have a creamy texture when cooked. Despite the name (*peruvian beans* in Spanish), they are native to Mexico. Yellow beans are uncommon in the United States due to a controversial patent issued in 1999 to John Proctor, who selected and named a strain of yellow bean from seeds he brought back from Mexico. U.S. Patent No. 5,894,079 (the Enola or yellow bean patent) granted POD-NERS, LLC., exclusive right to import and sell yellow beans in the United States from 1999 through 2008, when the patent was rejected after reexamination.

Production

In 2010, total world production of dry beans was 23 million metric tons, harvested from over 30 million hectares. World production of green beans in 2010 was 17.7 million ton, harvested from 15.1 million hectares.

Other Uses

Bean leaves have been used to trap bedbugs in houses. Microscopic hairs (trichomes) on the bean leaves entrap the insects.

From ancient times, beans were used as device in various methods of divination. Fortune-telling using beans is called favomancy.

Beets

Beta Vulgaris

Beta vulgaris (beet) is a plant in the Chenopodiaceae family which is now included in Amaranthaceae family. It has numerous cultivated varieties, the most well known of which is the root vegetable known as the beetroot or garden beet. However, other cultivated varieties include the leaf vegetable chard, as well as the root vegetable sugar beet, which is important in the production of table sugar, and mangelwurzel, which is a fodder crop. Three subspecies are typically recognised. All cultivated varieties fall into the subspecies *Beta vulgaris* subsp. *vulgaris*, while *Beta vulgaris* subsp. *maritima*, commonly known as the sea beet, is the wild ancestor of these, and is found throughout the Mediterranean, the Atlantic coast of Europe, the Near East, and India. A second wild subspecies, *Beta vulgaris* subsp. *adanensis*, occurs from Greece to Syria.

The roots are most commonly deep red-purple in colour, but come in a wide variety of other shades, including golden yellow and red-and-white striped.

Beta vulgaris is a herbaceous biennial or, rarely, perennial plant with leafy stems growing to 1–2 m tall. The leaves are heart-shaped, 5–20 cm long on wild plants (often much larger in cultivated plants). The flowers are produced in dense spikes; each flower is very small, 3–5 mm diametre, green or tinged reddish, with five petals; they are wind pollinated. The fruit is a cluster of hard nutlets.

Taxonomy

Figure: *Yellow-stemmed chard (with purple-leaved kale).*

The taxonomy of the various wild and cultivated races of beets has a long and complicated history. Mansfeld's Encyclopedia of Agricultural and Horticultural Crops following Letschert's 1993 treatment of *Beta*, section *Beta* recognizes the following taxa:

- *Beta all cultivated varieties of the beet, which are grown for their taproots, leaves, or swollen midribs.*
  - o *B. v.* ssp. *vulgaris* convar. *cicla* (leaf beets) - The leaf beet group has a long history dating to the second millennium BC. The first cultivated forms were believed to have been domesticated in the Mediterranean, but were introduced to the Middle East, India, and finally China by 850 AD. These were used as medicinal plants in Ancient Greece and Medieval Europe. Their popularity declined in Europe following the introduction of spinach.
    - – *B. v.* ssp. *v.* convar. *cicla.* var. *cicla* (spinach beet) - This variety is widely cultivated for its leaves, which are usually cooked like spinach. It can be found in many grocery stores around the world.
    - – *B. v.* ssp. *v.* convar. *cicla.* var. *flavescens* (chard) - Chard leaves have thick and fleshy midribs. Both the midribs and the leaf blades are used as vegetables, often in separate dishes. Some cultivars are also grown ornamentally for their coloured midribs. The thickened midribs are thought to have arisen from the spinach beet by mutation.
  - o *B. v.* ssp. *vulgaris* convar. *vulgaris* (tuberous beets) - This grouping contains all beets grown for their thickened tubers rather than their leaves.
    - – *B. v.* ssp. *v.* convar. *vulgaris* var. *crassa* (mangelwurzel) - This variety was developed in the 18th century for its tubers for use as a fodder crop.
    - – *B. v.* ssp. *v.* convar. *vulgaris* var. *altissima* (sugar beet) - The sugar beet is a major commercial crop due to its high concentrations of sucrose, which is extracted to produce table sugar. It was developed in Germany in the late 18th century after the roots of beets were found to contain sugar in 1747.
    - – *B. v.* ssp. *v.* convar. *vulgaris* var. *vulgaris* (beetroot or garden beet) - This is the red root vegetable that is most typically associated with the word 'beet'. It is especially popular in Eastern Europe where it is the main ingredient of borscht.

Ecology

Beets are a food plant by the larvae of a number of *Lepidoptera* species.

Uses

Food: Spinach beet leaves are eaten as a pot herb. Young leaves of the garden beet are sometimes used similarly. The midribs of Swiss chard are eaten boiled while the whole leaf blades are eaten as spinach beet.

In some parts of Africa, the whole leaf blades are usually prepared with the midribs as one dish.

The leaves and stems of young plants are steamed briefly and eaten as a vegetable; older leaves and stems are stir-fried and have a flavour resembling taro leaves.

The usually deep-red roots of garden beet are eaten boiled either as a cooked vegetable, or cold as a salad after cooking and adding oil and vinegar. A large proportion of the commercial production is processed into boiled and sterilised beets or into pickles. In Eastern Europe beet soup, such as cold borsch, is a popular dish. Yellow-coloured garden beets are grown on a very small scale for home consumption.

The consumption of beets causes pink urine in some people.

Jews traditionally eat beet on Rosh Hashana (New Year). Its Aramaic name sounds like the word for "remove" or "depart"; it is eaten with a prayer "that our enemies be removed".

Cultivation

Beets are cultivated for fodder (e.g. mangelwurzel), for sugar (the sugar beet), as a leaf vegetable (chard or "Bull's Blood"), or as a root vegetable ("beetroot", "table beet", or "garden beet").

"Blood Turnip" was once a common name for beet root cultivars for the garden. Examples include: Bastian's Blood Turnip, Dewing's Early Blood Turnip, Edmand Blood Turnip, and Will's Improved Blood Turnip.

The "earthy" taste of some beetroot cultivars comes from the presence of geosmin. Researchers have not yet answered whether beets produce geosmin themselves, or whether it is produced by symbiotic soil microbes living in the plant. Nevertheless, breeding programmes can produce cultivars with low geosmin levels yielding flavours more acceptable to consumers.

Beets are one of the most boron-intensive of modern crops, a dependency possibly introduced as an evolutionary response its preindustrial ancestor's constant exposure to sea spray; on commercial farms, a 60 tonne per hectare (26.8 ton/acre) harvest requires 600 grams of elemental boron per hectare (8.6 ounces/acre) for growth. A lack of boron causes the meristem and the shoot to languish, eventually leading to heart rot.

Red or Purple Colouring

***Figure:** A selection of different coloured beetroots.*

The colour of red/purple beetroot is due to a variety of betalain pigments, unlike most other red plants, such as red cabbage, which contain anthocyanin pigments. The composition of different betalain pigments can vary, giving breeds of beetroot which are yellow or other colours in addition to the familiar deep red. Some of the betalains in beets are betanin, isobetanin, probetanin, and neobetanin (the red to violet ones are known collectively as *betacyanin*). Other pigments contained in beet are indicaxanthin and vulgaxanthins (yellow to orange pigments known as *betaxanthins*). Indicaxanthin has been shown as a powerful protective antioxidant for thalassemia, as well as prevents the breakdown of alpha-tocopherol (Vitamin E).

Betacyanin in beetroot may cause red urine in some people who are unable to break it down. This is called beeturia.

The pigments are contained in cell vacuoles. Beetroot cells are quite unstable and will 'leak' when cut, heated, or when in contact with air or sunlight. This is why red beetroots leave a purple stain. Leaving the skin on when cooking, however, will maintain the integrity of the cells and therefore minimize leakage.

History

The sea beet, the ancestor of modern cultivated beets, prospered along the coast of the Mediterranean Sea. Beetroot remains have been excavated in the Third dynasty Saqqara pyramid at Thebes, Egypt, and four charred beetroots were found in the Neolithic site of Aartswoud in the Netherlands though it is difficult to determine whether these are domesticated or wild forms of *B. vulgaris*. Zohary and Hopf note that beetroot is "linguistically well identified." They state the earliest written mention of the beet comes from 8th century BC Mesopotamia. The Greek Peripatetic Theophrastus later describes the beet as similar to the radish, while Aristotle also mentions the plant. Available evidence, such as that provided by Aristotle and Theophrastus, suggests the leafy varieties of the beet were grown primarily for most of its history, though these lost much of their popularity much later following the introduction of spinach. The ancient Romans considered beets an important health food and an aphrodisiac.

Roman and Jewish literary sources suggest that by the 1st century BC the domestic beet was represented in the Mediterranean basin primarily by leafy forms like chard and spinach beet. Zohary and Hopf also argue that it is very probable that beetroot cultivars were also grown at the time, and some Roman recipes support this. Later English and German sources show that beetroots were commonly cultivated in Medieval Europe.

The Sugar Beet

Modern sugar beets date back to mid-18th century Silesia where the king of Prussia subsidised experiments aimed at processes for sugar extraction. In 1747 Andreas Marggraf isolated sugar from beetroots and found them at concentrations of 1.3-1.6%. He also demonstrated that sugar could be extracted from beets that was the same as that produced from sugarcane. His student, Franz Karl Achard, evaluated 23 varieties of mangelwurzel for sugar content and selected a local race from Halberstadt in modern-day Saxony-Anhalt, Germany. A man named Koppy and his son further selected from this race for white, conical tubers. The selection was named 'Weiße Schlesische Zuckerrübe', meaning white Silesian sugar beet, and boasted about a 6% sugar content. This selection is the progenitor of all modern sugar beets.

A royal decree led to the first factory devoted to sugar extraction from beetroots being opened in Kunern, Silesia (now Konary, Poland) in 1801. The Silesian sugar beet was soon introduced to France where Napoleon opened schools specifically for studying the plant. He also ordered that 28,000 hectares (69,000 acres) be devoted to growing the new sugar beet. This was in response to British blockades of cane sugar during the Napoleonic Wars, which ultimately stimulated the rapid growth of a European sugarbeet industry. By 1840 about 5% of the world's sugar was derived from sugar beets, and by 1880 this number had risen more than tenfold to over 50%. The sugar beet was introduced to North America after 1830 with the first commercial production starting in 1879 at a farm in Alvarado, California. The sugar beet was also introduced to Chile via German settlers around 1850.

It remains a widely cultivated commercial crop for producing table sugar.

Beetroot

The beetroot, also known in North America as the table beet, garden beet, red or golden beet, or informally simply as the beet, is any of the cultivated varieties of beet (*Beta vulgaris*) grown for their edible taproots, especially *B. vulgaris* L. subsp. *conditiva*. They are among the most commonly encountered varieties in North America, Central America, and Europe.

The usually deep red roots of beetroot are eaten either grilled, boiled, or roasted as a cooked vegetable, cold as a salad after cooking and adding oil and vinegar, or raw and shredded, either alone or combined with any salad vegetable. A large proportion of the commercial production is processed into boiled and sterilised beets or into pickles. In Eastern Europe, beet soup, such as borscht, is a popular dish. In Indian cuisine, chopped, cooked, spiced beet is a common side dish. Yellow-coloured beetroots are grown on a very small scale for home consumption.

The green, leafy portion of the beet is also edible. It is most commonly served boiled or steamed, in which case it has a taste and texture similar to spinach. Those selected should be bulbs that are unmarked, avoiding those with overly limp leaves or wrinkled skins, both of which are signs of dehydration.

Beetroot can be peeled, steamed, and then eaten warm with butter as a delicacy; cooked, pickled, and then eaten cold as a condiment; or peeled, shredded raw, and then eaten as a salad. Pickled beets are a traditional food of the American South. It is also common in Australia, New Zealand, and the United Arab Emirates for pickled beetroot to be served on a hamburger. A traditional Pennsylvania Dutch dish is red beet eggs. Hard-boiled eggs are refrigerated in the liquid left over from pickling beets and allowed to marinate until the eggs turn a deep pink-red colour.

In Poland, beet is combined with horseradish to form popular chrzanem, which is often added to a meal consisting of meat, potatoes and a salad.

When beet juice is used, it is most stable in foods with a low water activity, such as frozen novelties and fruit fillings. Betanins, obtained from the roots, are used industrially as red food colourants, e.g. to intensify the colour of tomato paste, sauces, desserts, jams and jellies, ice cream, sweets and breakfast cereals.

Beetroot can also be used to make wine.

Nutrients and Phytochemicals

Beetroot is an excellent source of folate and a good source of manganese, and contains betaines which may function to reduce the concentration of homocysteine, a homolog of the naturally occurring amino acid cysteine. High circulating levels of homocysteine may be harmful to blood vessels and thus contribute to the development of heart disease, stroke, or peripheral vascular disease. This hypothesis is controversial as it has not been established yet whether homocysteine itself is harmful, or whether it is just an indicator of increased risk for heart disease.

The red colour compound betanin is not broken down in the body, and in higher concentration may temporarily cause urine and stool to assume a reddish colour; in the case of urine this is called beeturia. This effect may cause distress and concern due to the visual similarity to hematuria (blood in the urine) or blood in the stool, but is completely harmless and will subside once the food is out of the system.

Preliminary Research

Basic research on rats as well as pilot studies on humans have shown betaine may protect against liver disease, particularly the accumulation of liver fat deposits caused by alcohol abuse, protein deficiency, or diabetes.

In preliminary research, beetroot juice reduced blood pressure in hypertensive individuals and so may have an effect on mechanisms of cardiovascular disease.

Dietary nitrate, such as that from consuming beets, may be a source for the biological messenger nitric oxide which induces the endothelium of arteries to signal smooth muscle, triggering vasodilation and increased blood flow.

One study found positive effects beetroot may have on human exercise performance, showing that distance runners ran 5% faster times after consuming baked beetroot.

A 2012 study tested the influence of dietary beet intake as a source of nitrate supplementation on resting heart rate and sustained apnea.

Drinking beetroot juice, containing approximately 5 mmol of nitrate, was found to reduce resting blood pressure by 2% and increase the maximum duration of apnea by 11% in experienced divers, relative to a control group receiving a placebo.

Other Uses

Betanin, obtained from the roots, is used industrially as red food colourant, to improve the colour and flavour of tomato paste, sauces, desserts, jams and jellies, ice cream, sweets, breakfast cereals, etc. Beetroot dye may also be used in ink.

Within older bulbs of beetroot, the colour is a deep crimson, and the flesh is much softer.

The juice has been used to make road salt more effective.

5

Leafy Vegetables

Leaf vegetables, also called potherbs, greens, vegetable greens, leafy greens or salad greens, are plant leaves eaten as a vegetable, sometimes accompanied by tender petioles and shoots. Although they come from a very wide variety of plants, most share a great deal with other leaf vegetables in nutrition and cooking methods.

Nearly one thousand species of plants with edible leaves are known. Leaf vegetables most often come from short-lived herbaceous plants such as lettuce and spinach. Woody plants whose leaves can be eaten as leaf vegetables include *Adansonia*, *Aralia*, *Moringa*, *Morus*, and *Toona* species.

The leaves of many fodder crops are also edible by humans, but usually only eaten under famine conditions. Examples include alfalfa, clover, and most grasses, including wheat and barley. These plants are often much more prolific than more traditional leaf vegetables, but exploitation of their rich nutrition is difficult, primarily because of their high fibre content. This obstacle can be overcome by further processing such as drying and grinding into powder or pulping and pressing for juice.

Leaf vegetables contain many typical plant nutrients, but since they are photosynthetic tissues, their vitamin K levels in relation to those of other fruits and vegetables, as well as other types of foods, is particularly notable. The reason is that phylloquinone, the most common form of the vitamin, is directly involved in photosynthesis. This causes leaf vegetables to be the primary food class that interacts significantly with the anticoagulant pharmaceutical warfarin.

During the first half of the 20th century, it was common for greengrocers to carry small bunches of herbs tied with a string to small green and red peppers, these bundles were called "potherbs."

Celery

Celery (*Apium graveolens* var. *dulce*) is a plant variety in the family Apiaceae, commonly used as a vegetable. The plant grows to 1 m (3.3 ft) tall. The leaves are pinnate to bipinnate with rhombic leaflets 3–6 cm long and 2–4 cm broad. The flowers are creamy-white, 2–3 mm in diametre, and are produced in dense compound umbels. The seeds are broad ovoid to globose, 1.5–2 mm long and wide.

First attested in English in 1664, the word "celery" derives from the French *céleri*, in turn from Italian *seleri*, the plural of *selero*, which comes from Late Latin *selinon*, the latinisation of the Greek óÝëéíïí (*selinon*), "parsley". The earliest attested form of the word is the Mycenaean Greek *se-ri-no*, written in Linear B syllabic script.

Taxonomy

Celery was described by Carl Linnaeus in Volume One of his *Species Plantarum* in 1753.

Cultivation

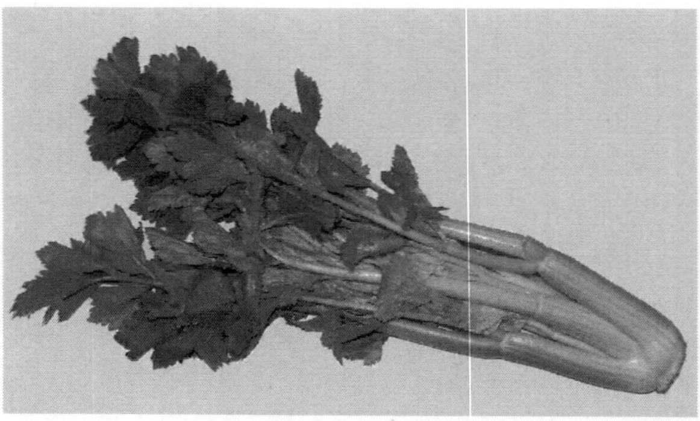

Figure: *Head of celery, sold as a vegetable. Usually only the stalks are eaten.*

In North America, commercial production of celery is dominated by the varieties called Pascal celery. Gardeners can grow a range of cultivars, many of which differ little from the wild species, mainly in having stouter leaf stems. They are ranged under two classes, white and red. The stalks grow in tight, straight, parallel bunches, and are typically marketed fresh that way, without roots and just a little green leaf remaining.

In Europe the dominant variety of celery most commonly available in trade is Celeriac (*Apium graveolens var. rapaceum*) grown for its hypocotyl forming a large bulb (commonly but incorrectly called celery root). The leaves are used as seasoning, and the stalks find only marginal use.

The wild form of celery is known as "smallage". It has a furrowed stalk with wedge-shaped leaves, the whole plant having a coarse, earthy taste, and a distinctive smell. The stalks are not usually eaten (except in soups or stews in French cuisine), but the leaves may be used in salads, and its seeds are those sold as a spice. With cultivation and blanching, the stalks lose their acidic qualities and assume the mild, sweetish, aromatic taste particular to celery as a salad plant.

The plants are raised from seed, sown either in a hot bed or in the open garden according to the season of the year, and, after one or two thinnings and transplantings, they are, on attaining a height of 15–20 cm, planted out in deep trenches for convenience of blanching, which is effected by earthing up to exclude light from the stems.

In the past, celery was grown as a vegetable for winter and early spring; it was perceived as a cleansing tonic, welcomed to counter the salt-sickness of a winter diet. By the 19th century, the season for celery had been extended, to last from the beginning of September to late in April.

Harvesting and Storage

Harvesting occurs when the average size of celery in a field is marketable; due to extremely uniform crop growth, fields are harvested only once. The petioles and leaves are removed and harvested; celery is packed by size and quality (determined by colour, shape, straightness and thickness of petiole, stalk and midrib length and absence of disease, cracks, splits, insect damage and rot). Under optimal conditions, celery can be stored for up to seven weeks between 0 to 2 °C (32 to 36 °F). Inner stalks may continue growing if kept at temperatures above 0 °C (32 °F). Freshly cut petioles of celery are prone to decay, which can be prevented or reduced through the use of sharp blades during processing, gentle handling, and proper sanitation.

In the past, restaurants used to store it in a container of water with powdered vegetable preservative; however, the sulfites in the preservative caused allergic reactions in some people. In 1986, the U.S. Food and Drug Administration banned the use of sulfites on fruits and vegetables intended to be eaten raw.

Uses

Celery is used around the world as a vegetable for the crisp petiole (leaf stalk). The leaves are strongly flavoured and are used less often, either as a flavouring in soups and stews or as a dried herb.

In temperate countries, celery is also grown for its seeds. Actually very small fruit, these "seeds" yield a valuable volatile oil used in the perfume and pharmaceutical industries. They also contain an organic compound called apiol. Celery seeds can be used as flavouring or spice, either as whole seeds or ground and mixed with salt, as celery salt. Celery salt can also be made from an extract of the roots, or using dried leaves. Celery salt is used as a seasoning, in cocktails (notably to enhance the flavour of Bloody Mary cocktails), on the Chicago-style hot dog, and in Old Bay Seasoning.

Celery, onions, and bell peppers are the "holy trinity" of Louisiana Creole and Cajun cuisine. Celery, onions, and carrots make up the French mirepoix, often used as a base for sauces and soups. Celery is a staple in many soups, such as chicken noodle soup.

Medicine

The use of celery seed in pills for relieving pain was described by Aulus Cornelius Celsus around AD 30. Celery seeds contain a compound, 3-*n*-butylphthalide, that has been demonstrated to lower blood pressure in rats.

Bergapten in the seeds can increase photosensitivity, so the use of essential oil externally in bright sunshine should be avoided. The oil and large doses of seeds should be avoided during pregnancy, as they can act as a uterine stimulant. Seeds intended for cultivation are not suitable for eating as they are often treated with fungicides.

Nutrition

Celery is used in weight-loss diets, where it provides low-calorie dietary fibre bulk. Celery is often incorrectly purported to be a "negative-calorie food" based on the idea that the body will burn more calories during the digestion of the food than the body can extract from the food itself. However, the fact that the body uses very small amounts of energy in digestion compared to what can be extracted even from a low-calorie food like celery disproves this hypothesis.

Allergies

Celery is among a small group of foods (headed by peanuts) that appear to provoke the most severe allergic reactions; for people with celery allergy, exposure can cause potentially fatal anaphylactic shock. The allergen does not appear to be destroyed at cooking temperatures. Celery root—commonly eaten as celeriac, or put into drinks—is known to contain more allergen than the stalk. Seeds contain the highest levels of allergen content. Exercise-induced anaphylaxis may be exacerbated.

An allergic reaction also may be triggered by eating foods that have been processed with machines that have previously processed celery, making avoiding such foods difficult. In contrast with peanut allergy being most prevalent in the US, celery allergy is most prevalent in Central Europe. In the European Union, foods that contain or may contain celery, even in trace amounts, must be clearly marked as such.

Chemistry

Polyacetylenes can be found in Apiaceae vegetables like celery, and their extracts show cytotoxic activities.

History

Daniel Zohary and Maria Hopf note that celery leaves and inflorescences were part of the garlands found in the tomb of pharaoh Tutankhamun (died 1323 BC), and celery mericarps dated to the seventh century BC were recovered in the Heraion of Samos. However, they note "since *A. graveolens* grows wild in these areas, it is hard to decide whether these remains represent wild or cultivated forms." Only by classical times is it certain that celery was cultivated.

M. Fragiska mentions an archeological find of celery dating to the 9th century BC, at Kastanas; however, the literary evidence for ancient Greece is far more abundant. In Homer's *Iliad*, the horses of the Myrmidons graze on wild celery that grows in the marshes of Troy, and in *Odyssey*, there is mention of the meadows of violet and wild celery surrounding the cave of Calypso.

Cultural Depictions

A chthonian symbol among the ancient Greeks, celery was said to have sprouted from the blood of Kadmilos, father of the Cabeiri, chthonian divinities celebrated in Samothrace, Lemnos, and Thebes. The spicy odour and dark leaf colour encouraged this association with the cult of death. In classical Greece, celery leaves were used as garlands for the dead, and the wreaths of the winners at the Isthmian Games were first made of celery before being replaced by crowns made of pine. According to Pliny the Elder in Achaea, the garland worn by the winners of the sacred Nemean Games was also made of celery. The Ancient Greek colony of Selinous (Greek: Óåëéíïæò, *Selinous*), on Sicily, was named after wild parsley that grew abundantly there; Selinountian coins depicted a parsley leaf as the symbol of the city.

The name "celery" retraces the plant's route of successive adoption in European cooking, as the English "celery" (1664) is derived from the French *céleri* coming from the Lombard term, *seleri*, from the Latin *selinon*, borrowed from Greek.

Celery's late arrival in the English kitchen is an end-product of the long tradition of seed selection needed to reduce the sap's bitterness and increase its sugars. By 1699, John Evelyn could recommend it in his *Acetaria. A Discourse of Sallets*: "Sellery, apium Italicum, (and of the Petroseline Family) was formerly a stranger with us (nor very long since in Italy) is an hot and more generous sort of Macedonian Persley or Smallage...and for its high and grateful Taste is ever plac'd in the middle of the Grand Sallet, at our Great Men's tables, and Praetors feasts, as the Grace of the whole Board".

Celery has made a surprising appearance in football folklore. Supporters of English Premier League team Chelsea and Football League team Gillingham regularly sing songs about the vegetable and are famed for throwing celery during matches. This has also given rise to the "Chelsea Cocktail", a pint of Guinness garnished with a stick of celery.

In the television programme *Doctor Who*, the Fifth incarnation of The Doctor (played by Peter Davison), was noted for wearing a stalk of celery on his lapel, claiming it at one point to be an excellent restorative, though the human olfactory sense was comparatively weak.

Celery Growing Problems

Celery requires plenty of moisture, fertile soil, and a long, cool growing season. You can start celery from seed, but germination is slow, about four weeks. Sow seeds indoors at least 10 weeks before seedlings are to be set out. A better plan might be to purchase celery starts when you are ready to plant.

Even with moisture, fertile soil, and the right climate for growing, celery is seldom trouble-free. Celery is susceptible to many pests and diseases. Here is a list of possible celery growing problems matched with cures and controls:

Common celery growing problems with cures and controls:

Seeds rot or seedlings collapse with dark water-soaked stems as soon as they appear. Damping off is a fungus that lives in the soil, particularly where humidity is high. Do not plant in cold, moist soil. Make sure soil is well drained.

Seedlings stunted, plants appear stunted; roots appear to have knots or beads. Root-knot nematodes are microscopic worm-like animals that live in the film of water that coat soil particles; some are pests, some are not. Root-knot nematodes feed in the roots and stunt plant growth; they are most common in sandy soils. Rotate crops. Solarize the soil with clear plastic in mid-summer.

Plants produce lots of leaves but not stalks; growth is slow. Sudden temperature fluctuations during early growth. Protect young plants from cold; use horticultural cloth or cloches when temperatures are low. Don't plant too early.

Bolting; plants flower and go to seed. Celery will bolt prematurely if plants are exposed to too many days with temperatures below 55°F. Protect young plants from cold; use horticultural cloth or cloches when temperatures are low. Don't plant too early.

Leaves turn yellow and then brown from the bottom up; plant loses vigor; plants appear stunted; worms bore into roots. Wireworms are the soil-dwelling larvae of click beetles; they look like wiry-jointed worms. Check soil before planting; flood the soil if wireworms are present. Remove infested plants and surrounding soil.

Young stems chewed. Young earwigs feed on plant shoots and eat holes in foliage. Most often the damage is tolerable and the infestation is light. Heavy infestation use traps of rolled wet paper or old flowerpots stuffed with paper to catch earwigs at night. Dump them in soapy water. Keep garden free of plant debris. Spray with hot pepper and garlic repellent.

Leaves curl under and become deformed and yellowish. Aphids are tiny, oval, and yellowish to greenish pear-shaped insects that colonize on the undersides of leaves. They leave behind sticky excrement called honeydew which can turn into a black sooty mold. Remove with a blast of water. Use insecticidal soap. Aluminum foil mulch will leave aphids disoriented.

Tunnels or blotches in leaves. Leafminer larvae tunnel inside leaves. Destroy infected leaves and caterpillars. Cultivate the garden to destroy larvae and keep adult flies from laying eggs. Cover crops with floating row covers to exclude flies.

Leaves folded under and webbed; holes chewed in leaves and stalks. Celery leaftier is a pale green caterpillar with a white stripe down its back that grows to ¾-inches long; the adult moth is small brown with dark wavy lines in its wings. Hand pick and destroy. Remove leaves which shelter caterpillar. Use Bacillus thuringiensis.

Irregular small holes eaten in leaves. Cabbage looper is a light green caterpillar with yellow stripes running down the back; loops as it walks. Keep garden clean of debris where adult brownish night-flying moth can lay eggs. Cover plants with spun polyester to exclude moths. Pick loppers off by hand. Use Bacillus thuringiensis. Dust with Sevin or rotenone.

Irregular holes in foliage and stems. Celeryworm is the grown and white larva of the black swallow-tail butterfly. Usually few are found. Pick off by hand. Spray with Bacillus thuringiensis (Dipel). Spray or dust with Sevin, pyrethrum, rotenone.

Leaves are eaten and plants are partially defoliated. Blister beetles and tomato hornworms eat leaves. Handpick insects and destroy. Keep the garden weeds and debris. Cultivate in spring to kill larvae and interrupt the life cycle. Pick off beetles by hand. Spray or dust with Sevin or use a pyrethrum or rotenone spray.

Large holes eaten in leave; trails of sliver slime. Snails and slugs prefer cool temperatures. Hand pick and destroy; place protective borders of sand, lime or wood ashes around plants. Mulch with wood shavings or oak leaves.

Tips and leaf margins become streaked and look scorched. Magnesium deficiency. Have the soil tested. Use magnesium chelates. Plant resistant varieties: Emerson Pascal, Utah 52-75.

Brown spots on leaves and stems; plants become stunted and die. Late blight of celery is caused by Septoria fungus a soil and seedborne disease triggered by heat and humidity following a rainy period. Keep the garden clean and free of weeds. Remove infected plants. Improve soil drainage.

Yellow spots on outer leaves enlarge to become gray-brown streaks. Celery early blight or Cercospora leaf spot is a fungal disease spread by heavy rainfall and warm temperatures. Keep weeds down in the garden area; they harbor fungal spores. Avoid overhead watering.

Plant turns pale green, yellows beginning on one side, yellowing spreads; stems rot; plant wilts. Fusarium wilt or stem rot is soilborne fungus that infects plant vascular tissue, usually where the soil is warm. Plant in well-drained soil. Rotate crops. Remove and destroy infected plants. Grow resistant varieties: Cornell No. 19.

Mottled green and yellow leaves; zigzag bands may develop, distorted leaves. Mosaic virus has no cure; it is spread from plant to plant by aphids and leafhoppers. Plant resistant varieties. Remove diseased plants. Remove broadleaf weeds that serve as virus reservoir.

Zigzag paths into crown and heart of celery. White with brown, legless grubs are the larvae of the carrot weevil, a dark crown to coppery, hard-shelled weevil to 1/5 inch long. Stunted plants Carrot weevil larvae are white grubs with dark heads that feed at the crown of celery.

Twisted, brittle stalks; plants yellowed and stunted. Aster yellows is a mycoplasma disease spread by leafhoppers. Remove infected plants. Control leafhopper. Keep the garden free of weeds which can harbor disease.

Inner stalks and leaves die at tips and turn brown or black. Blackheart associated with soil calcium deficiency. Calcium deficiency can inhibit uptake of water into plant. Increase watering and mulch to conserve soil moisture. Test soil; add gypsum or limestone if soil is calcium deficient. Maintain soil pH between 6.5 and 8.0. Plant resistant celery varieties: Cornell 19, Emerald, Emerson, Golden Pascal.

Celery stems crack crosswise; stems are stiff and brittle; leaf edges may be streaked and brown. Boron deficiency, often found in alkaline soils. Test soil. If deficient, add 2 ounces of borax per 30 square yards. Plant resistant varieties of celery: Golden Self-Blanching, Dwarf Golden Self-Blanching, Giant Pascal, Utah 52-70.

Central leaves turn dark. Similar to blackheart, but this is the work of the tarnished plant bug a greenish yellow to brown bug to ¼-inch long wit yellow triangle marks at the end of each wing. The tarnished plant bug sucks sap from plant causing tissue to break down; leaves at the centre of the plant turn gray and black while outer stalks appear healthy.

Water-soaked spots on stalks with cottony pink mold at base. Pink rot fungus grows at soil surface level in cool, wet conditions. Remove and destroy infected plants. Keep garden free of weeds where fungal spores can rest. Stalks tough, bitter. Possibly: stalks over-mature; high temperatures; dry soil; poor fertility. Harvest when tender. Plant so that celery comes to maturity in cool weather. Keep plants evenly moist throughout growing season. Side dress feed plants with aged compost.

Chaulai (Amaranth)

Amaranthus, collectively known as amaranth, is a cosmopolitan genus of annual or short-lived perennial plants. Catkin-like cymes of densely packed flowers grow in summer or autumn. Approximately 60 species are recognized, with inflorescences and foliage ranging from purple and red to green or gold. Members of this genus share many characteristics and uses with members of the closely related genus *Celosia*.

Although several species are often considered weeds, people around the world value amaranths as leaf vegetables, cereals, and ornamental plants.

"Amaranth" derives from Greek ìÜñáíôïò (*amarantos*), "unfading," with the Greek word for "flower," íèïò (*anthos*), factoring into the word's development as "amaranth." The more accurate "amarant" is an archaic variant.

Amaranthus shows a wide variety of morphological diversity among and even within certain species. Although the family (Amaranthaceae) is distinctive, the genus has few distinguishing characters among the 70 species included. This complicates taxonomy and *Amaranthus* has generally been considered among systematists as a "difficult" genus.

Formerly, Sauer (1955) classified the genus into two subgenera, differentiating only between monoecious and dioecious species: *Acnida* (L.) Aellen ex K.R. Robertson and *Amaranthus*. Although this classification was widely accepted, further infrageneric classification was (and still is) needed to differentiate this widely diverse group.

Currently, *Amaranthus* includes three recognized subgenera and 70 species, although species numbers are questionable due to hybridization and species concepts. Infrageneric classification focuses on inflorescence, flower characters and whether a species is monoecious/dioecious, as in the Sauer (1955) suggested classification. A modified infrageneric classification of *Amaranthus* was published by Mosyakin & Robertson (1996) and includes three subgenera: *Acnida*, *Amaranthus*, and *Albersia*. The taxonomy is further differentiated by sections within each of the subgenera.

History

Known to the Aztecs as huauhtli, it is thought to have represented up to 80% of their caloric consumption before the conquest. Another important use of amaranth throughout Mesoamerica was to prepare ritual drinks and foods. To this day, amaranth grains are toasted much like popcorn and mixed with honey, molasses or chocolate to make a treat called *alegría*, meaning "joy" in Spanish. Diego Duran described the festivities for Huitzilopochtli, whose name means "hummingbird of the left side" or "left-handed hummingbird". (Real hummingbirds feed on amaranth flowers.)

The Aztec month of Panquetzaliztli (7 December to 26 December) was dedicated to Huitzilopochtli. People decorated their homes and trees with paper flags; there were ritual races, processions, dances, songs, prayers, and finally human sacrifices. This was one of the more important Aztec festivals, and the people prepared for the whole month. They fasted or ate very little; a statue of the god was made out of amaranth (*huautli*) seeds and honey, and at the end of the month, it was cut into small pieces so everybody could eat a little piece of the god. After the Spanish conquest, cultivation of amaranth was outlawed, while some of the festivities were subsumed into the Christmas celebration.

Because of its importance as a symbol of indigenous culture, its gluten-free palatability, ease of cooking, and a protein that is particularly well-suited to human nutritional needs, interest in grain amaranth (especially *A. cruentus* and *A. hypochondriacus*) revived in the 1970s. It was recovered in Mexico from wild varieties and is now commercially cultivated. It is a popular snack sold in Mexico, sometimes mixed with chocolate or puffed rice, and its use has spread to Europe and parts of North America. Amaranth and quinoa are called pseudograins because of their flavour and cooking similarities to grains.

Amaranth Seed

Several species are raised for amaranth "grain" in Asia and the Americas.

Ancient amaranth grains still used to this day include the three species, *Amaranthus caudatus*, *Amaranthus cruentus*, and *Amaranthus hypochondriacus*. Although amaranth was cultivated on a large scale in ancient Mexico, Guatemala, and Peru, nowadays it is only cultivated on a small scale there, along with India, China, Nepal, and other tropical countries; thus, there is potential for further cultivation in those countries, as well as in the U.S. In a 1977 article in *Science*, amaranth was described as "the crop of the future." It has been proposed as an inexpensive native crop that could be cultivated by indigenous people in rural areas for several reasons:

1. It is easily harvested.
2. Its seeds are a good source of protein. Compared to other grains, amaranth is unusually rich in the essential amino acid lysine. Common grains such as wheat and corn are comparatively rich in amino acids that amaranth lacks; thus, amaranth and other grains can complement each other.
3. The seeds of *Amaranthus* species contain about thirty percent more protein than cereals like rice, sorghum and rye. In cooked and edible forms, amaranth is competitive with wheat germ and oats - higher in some nutrients, lower in others.
4. It is easy to cook. As befits its weedy life history, amaranth grains grow very rapidly and their large seedheads can weigh up to 1 kilogram and contain a half-million seeds in three species of amaranth.

Amaranth Seed Flour

Amaranth seed flour has been evaluated as an additive to wheat flour by food specialists. To determine palatability, different levels of amaranth grain flour were mixed with the wheat flour and baking ingredients (1% salt, 2.5% fat, 1.5% yeast, 10% sugar and 52–74% water), fermented, molded, pan-proofed and baked. The baked products were evaluated for loaf volume, moisture content, colour, odour, taste and texture. The amaranth containing products were then compared with bread made from 100% wheat flour.

The loaf volume decreased by 40% and the moisture content increased from 22 to 42% with increase in amaranth grain flour. The study found that the sensory scores of the taste, odour colour and texture decreased with increasing amounts of amaranth. Generally, above 15% amaranth grain flour, there were significant differences in the evaluated sensory qualities and the high amaranth-containing product was found to be of unacceptable palatability to the population sample that evaluated the baked products.

Leaves, Roots, and Stems

Amaranth species are cultivated and consumed as a leaf vegetable in many parts of the world. There are four species of *Amaranthus* documented as cultivated vegetables in eastern Asia: *Amaranthus cruentus*, *Amaranthus blitum*, *Amaranthus dubius*, and *Amaranthus tricolour*.

In Indonesia and Malaysia, leaf amaranth is called *bayam*. In the Philippines, the Ilocano word for the plant is "kalunay"; the Tagalog word for the plant is *kilitis* or "kulitis". In the state of Uttar Pradesh and Bihar in India, it is called Chaulai and is a popular green leafy vegetable (referred to in the class of vegetable preparations called saag). It is called Chua in Kumaun area of Uttarakhand, where it is a popular red-green vegetable. In Karnataka state in India, it is called Harive. It is used to prepare curries like Hulee, palya, Majjigay-hulee and so on. In the state of Kerala, it is called 'Cheera' and is consumed by stir-frying the leaves with spices and red chillies to make 'Cheera Thoran'. In Maharashtra, it is called "Shravani Maath" (literally grown in month of *Shravan*) and it is available in both red and white colour. In Orissa, it is called "Khada saga", it is used to prepare 'Saga Bhaja', in which the leaf is fried with chillies and onions.

The root of mature amaranth is a popular vegetable. It is white and cooked with tomatoes or tamarind gravy. It has a milky taste and is alkaline.

In China, the leaves and stems are used as a stir-fry vegetable, or in soups, and called (Mandarin Pinyin: xiàncài; Cantonese Jyutping: jin6 coi3) with variations in various dialects). Amaranth greens are believed to help enhance eyesight. In Vietnam, it is called *rau dÁn* and is used to make soup. There are two species popular as edible vegetable in Vietnam: *dÁn ðĬ*- amaranthus tricolour and *dÁn cõm* or *dÁn tr¯ng*- amaranthus viridis.

A traditional food plant in Africa, amaranth has the potential to improve nutrition, boost food security, foster rural development and support sustainable land care. In East Africa, amaranth leaf is known in chewa as *bonongwe*, and in Swahili as *mchicha*, as *terere* in Kikuyu, Meru and Embu; and as *telele* in Kamba. In Bantu regions of Uganda it is known as *doodo*. It is recommended by some doctors for people having low red blood cell count. It is also known among the Kalenjin as a drought crop (*chepkerta*). In Lingala (spoken in the Congo), it is known as *l[ngal[nga* or *bít[kut[ku*. In Nigeria, it is a common vegetable and goes with all Nigerian starch dishes. It is known in Yoruba as *efo tete* or *arowo jeja* (meaning "we have money left over for fish"). In the Caribbean, the leaves are called *bhaji* in Trinidad and *callaloo* in Jamaica, and are stewed with onions, garlic and tomatoes, or sometimes used in a soup called pepperpot soup.

In Greece, green amaranth (*Amaranthus viridis*) is a popular dish and is called *vlita* or *vleeta*. It is boiled, then served with olive oil and vinegar like a salad, usually alongside fried fish. Greeks stop harvesting the plant (which usually grows wild) when it starts to bloom at the end of August.

In Sri Lanka, it is called "koora thampala". Sri Lankans cook it and eat it with rice. Fiji Indians call it *choraiya bhaji*.

Dyes

The flowers of the 'Hopi Red Dye' amaranth were used by the Hopi (a tribe in the western United States) as the source of a deep red dye. There is also a synthetic dye that has been named "amaranth" for its similarity in colour to the natural amaranth pigments known as betalains. This synthetic dye is also known as *Red No. 2* in North America and E123 in the European Union.

Ornamentals

The genus also contains several well-known ornamental plants, such as *Amaranthus caudatus* (love-lies-bleeding), a native of India and a vigorous, hardy annual with dark purplish flowers crowded in handsome drooping spikes. Another Indian annual, *A. hypochondriacus* (prince's feather), has deeply veined lance-shaped leaves, purple on the under face, and deep crimson flowers densely packed on erect spikes.

Amaranths are recorded as food plants for some Lepidoptera (butterfly and moth) species including the nutmeg moth and various case-bearer moths of the genus *Coleophora*: *C. amaranthella*, *C. enchorda* (feeds exclusively on *Amaranthus*), *C. immortalis* (feeds exclusively on *Amaranthus*), *C. lineapulvella* and *C. versurella* (recorded on *A. spinosus*).

Nutritional Value

Amaranth greens are a common leaf vegetable throughout the tropics and in many warm temperate regions.

Cooked amaranth leaves are a good source of vitamin A, vitamin C, and folate; they are also a complementing source of other vitamins such as thiamine, niacin, and riboflavin, plus some dietary minerals including calcium, iron, potassium, zinc, copper, and manganese. Cooked amaranth grains are a complementing source of thiamine, niacin, riboflavin, and folate, and dietary minerals including calcium, iron, magnesium, phosphorus, zinc, copper, and manganese - comparable to common grains such as wheat germ, oats and others.

Amaranth seeds contain lysine, an essential amino acid, limited in other grains or plant sources. Most fruits and vegetables do not contain a complete set of amino acids, and thus different sources of protein must be used. Amaranth too is limited in some essential amino acids, such as leucine and threonine. Amaranth seeds are therefore a promising complement to common grains such as wheat germ, oats, and corn because these common grains are abundant sources of essential amino acids found to be limited in amaranth.

Amaranth may be a promising source of protein to those who are gluten sensitive, because unlike the protein found in grains such as wheat and rye, its protein does not contain gluten. According to a 2007 report, amaranth compares well in nutrient content with gluten-free vegetarian options such as buckwheat, corn, millet, wild rice, oats and quinoa.

Several studies have shown that like oats, amaranth seed or oil may be of benefit for those with hypertension and cardiovascular disease; regular consumption reduces blood pressure and cholesterol levels, while improving antioxidant status and some immune parametres. While the active ingredient in oats appears to be water-soluble fibre, amaranth appears to lower cholesterol via its content of plant stanols and squalene.

Amaranth remains an active area of scientific research for both human nutritional needs and foraging applications. Over 100 scientific studies suggest a somewhat conflicting picture on possible anti-nutritional and toxic factors in amaranth, more so in some particular strains of amaranth. Lehmann, in a review article, identifies some of these reported anti-nutritional factors in amaranth to be phenolics, saponins, tannins, phytic acid, oxalates, protease inhibitors, nitrates, polyphenols and phytohemagglutinins.

Of these, oxalates and nitrates are of more concern when amaranth grain is used in foraging applications. Some studies suggest thermal processing of amaranth, particularly in moist environment, prior to its preparation in food and human consumption may be a promising way to reduce the adverse effects of amaranth's anti-nutritional and toxic factors.

Ecology

Not all amaranth plants are cultivated. Most of the species from *Amaranthus* are summer annual weeds and are commonly referred to as pigweeds. These species have an extended period of germination, rapid growth, and high rates of seed production, and have been causing problems for farmers since the mid-1990s. This is partially due to the reduction in tillage, reduction in herbicidal use and the evolution of herbicidal resistance in several species where herbicides have been applied more often. The following 9 species of *Amaranthus* are considered invasive and noxious weeds in the U.S and Canada: *A. albus*, *A. blitoides*, *A. hybridus*, *A. palmeri*, *A. powellii*, *A. retroflexus*, *A. spinosus*, *A. tuberculatus*, and *A. viridis*.

A new herbicide-resistant strain of *Amaranthus palmeri* has appeared; it is glyphosate-resistant and so cannot be killed by herbicides using the chemical. Also, this plant can survive in tough conditions.This could be of particular concern to cotton farmers using glyphosate-resistant cotton. The species *Amaranthus palmeri* (Palmer amaranth) causes the greatest reduction in soybean yields and has the potential to reduce yields by 17-68% in field experiments.

Palmer amaranth is among the "top five most troublesome weeds" in the southeast of the United States and has already evolved resistances to dinitroaniline herbicides and acetolactate synthase inhibitors. This makes the proper identification of *Amaranthus* species at the seedling stage essential for agriculturalists. Proper weed control needs to be applied before the species successfully colonizes in the crop field and causes significant yield reductions.

Seed Saving

There are a multitude of varieties which cross with one another very easily. Some species have been found to cross with one another e.g. *Amaranthus caudatus* and *Amaranthus hypochondriacus*. For most types, flowering occurs as the days become shorter.

Being wind-pollinated, they will cross with one another if less than 400 metres apart at flowering time. If the seed is to be used for planting, roguing is necessary to remove inferior individuals before they can flower and pollinate better plants.

The seed heads mature gradually from bottom to top, requiring harvesters to be selective when choosing plants for seed harvesting. Seed harvest is maximized by shaking the near-mature seed heads into a paper bag or onto a canvas.

In large growing areas the heads are cut all at once when most of the seeds are ripe. Once the heads have fully ripened, they tend to drop their seeds, so harvesting is done just before this point.

Heads are then dried for a week and threshed with gloved hands or feet on canvas as the chaff is somewhat prickly. Care is required not to lose the seeds when winnowing because the chaff and seeds are of similar size and the seeds are of a light weight. Heaping uncleaned seeds in a bowl and tossing them will concentrate the light debris on the top, and it can then then blown away. The process is repeated until only seeds remain.

Bathua

Chenopodium album is a fast-growing weedy annual plant in the genus *Chenopodium*.

Though cultivated in some regions, the plant is elsewhere considered a weed. Common names include lamb's quarters, melde, goosefoot and fat-hen, though the latter two are also applied to other species of the genus *Chenopodium*, for which reason it is often distinguished as white goosefoot. It is sometimes also called pigweed.

However, Pigweed is also a name for a few weeds in the family: Amaranthaceae, The Name pigweed is used for *Amaranthus albus*, Redroot pigweed and others.

Chenopodium album is extensively cultivated and consumed in Northern India as a food crop, and in English texts it may be called by its Hindi name *bathua* or *bathuwa*. It is called Pappukura in Telugu, Paruppukkirai in Tamil, Kaduoma in Kannada, Vastuccira in Malayalam, and Chakvit in Konkani.

Figure: *Young* Chenopodium album

Distribution

Its native range is obscure due to extensive cultivation, but includes most of Europe, from where Linnaeus described the species in 1753. Plants native in eastern Asia are included under *C. album*, but often differ from European specimens. It is widely introduced elsewhere, e.g. Africa, Australasia, North America, and Oceania, and now occurs almost everywhere in soils rich in nitrogen, especially on wasteland.

Botany

It tends to grow upright at first, reaching heights of 10–150 cm (rarely to 3 m), but typically becomes recumbent after flowering (due to the weight of the foliage and seeds) unless supported by other plants. The leaves are alternate and can be varied in appearance. The first leaves, near the base of the plant, are toothed and roughly diamond-shaped, 3–7 cm long and 3–6 cm broad. The leaves on the upper part of the flowering stems are entire and lanceolate-rhomboid, 1–5 cm long and 0.4–2 cm broad; they are waxy-coated, unwettable and mealy in appearance, with a whitish coat on the underside. The small flowers are radially symmetrical and grow in small cymes on a dense branched inflorescence 10–40 cm long.

Taxonomy

Chenopodium album has a very complex taxonomy and has been divided in numerous microspecies, subspecies and varieties, but it is difficult to differentiate between them. The following infraspecific taxa are accepted by the *Flora Europaea*:

- *Chenopodium album* subsp. *album*
- *Chenopodium album* subsp. *striatum* (Krašan) Murr
- *Chenopodium album* var. *reticulatum* (Aellen) Uotila

Published names and synonyms include *C. album* var. *microphyllum*, *C. album* var. *stevensii*, *C. acerifolium*, *C. centrorubrum*, *C. giganteum*, *C. jenissejense*, *C. lanceolatum*, *C. pedunculare* and *C. probstii*.

It also hybridises readily with several other *Chenopodium* species, including *C. berlandieri*, *C. ficifolium*, *C. opulifolium*, *C. strictum* and *C. suecicum*.

Cultivation

Regions where Cultivated: The species are cultivated as a grain or vegetable crop (such as in lieu of spinach), as well as animal feed in Asia and Africa, whereas in Europe and North America, it is commonly regarded as a weed in places such as potato fields.

Potential Impact on Conventional Crops

It is one of the more robust and competitive weeds, capable of producing crop losses of up to 13% in corn, 25% in soybeans, and 48% in sugar beets at an average plant distribution. It may be controlled by dark tillage, rotary hoeing, or flaming when the plants are small. Crop rotation of small grains will suppress an infestation. It is easily controlled with a number of pre-emergence herbicides. Its pollen may contribute to hay fever-like allergies.

Beneficial Use in Ecological Pest Control

Chenopodium album is vulnerable to leaf miners, making it a useful trap crop as a companion plant. Growing near other plants, it attracts leaf miners which might otherwise have attacked the crop to be protected. It is a host plant for the beet leafhopper, an insect which transmits curly top virus to beet crops.

Uses and Consumption

Food: The leaves and young shoots may be eaten as a leaf vegetable, either steamed in its entirety, or cooked like spinach, but should be eaten in moderation due to high levels of oxalic acid. Each plant produces tens of thousands of black seeds. These are high in protein, vitamin A, calcium, phosphorus, and potassium. Quinoa, a closely related species, is grown specifically for its seeds. The Zuni people cook the young plants' greens.

Archaeologists analysing carbonized plant remains found in storage pits and ovens at Iron Age, Viking Age, and Roman sites in Europe have found its seeds mixed with conventional grains and even inside the stomachs of Danish bog bodies.

In India, the plant is popularly called *bathua* and found abundantly in the winter season. The leaves and young shoots of this plant are used in dishes such as soups, curries, and paratha-stuffed breads, especially popular in Punjab. The seeds or grains are used in *phambra* or *laafi*, gruel-type dishes in Himachal Pradesh, and in mildly alcoholic fermented beverages such as *soora* and *ghanti*.

Animal Feed: As some of the common names suggest, it is also used as feed (both the leaves and the seeds) for chickens and other poultry.

Methi

Fenugreek *Trigonella foenum-graecum*) is an annual plant in the family Fabaceae with leaves consisting of three small obovate to oblong leaflets. It is cultivated worldwide as a semi-arid crop, and its seeds are a common ingredient in dishes from the Indian Subcontinent.

It is believed fenugreek was brought into cultivation in the Near East. While Zohary and Hopf are uncertain which wild strain of the genus *Trigonella* gave rise to domesticated fenugreek, charred fenugreek seeds have been recovered from Tell Halal, Iraq, (carbon dated to 4000 BC) and Bronze Age levels of Lachish and desiccated seeds from the tomb of Tutankhamen. Cato the Elder lists fenugreek with clover and vetch as crops grown to feed cattle.

Production

Major fenugreek-producing countries are Afghanistan, Pakistan, India, Iran, Nepal, Bangladesh, Argentina, Egypt, France, Spain, Turkey and Morocco. The largest producer is India, where the major producing states are Rajasthan, Gujarat, Uttarakhand, Uttar Pradesh, Madhya Pradesh, Maharashtra, Haryana, and Punjab. Rajasthan accounts for over 80% of India's output.

Use

Cuisine: Fenugreek is used as an herb (dried or fresh leaves), spice (seeds), and vegetable (fresh leaves, sprouts, and microgreens). Sotolon is the chemical responsible for fenugreek's distinctive sweet smell.

Cuboid-shaped, yellow-to-amber coloured fenugreek seeds are frequently encountered in the cuisines of the Indian subcontinent, used both whole and powdered in the preparation of pickles, vegetable dishes, daals, and spice mixes such as panch phoron and sambar powder. They are often roasted to reduce bitterness and enhance flavour.

Fresh fenugreek leaves are an ingredient in some Indian curries. Sprouted seeds and microgreens are used in salads. When harvested as microgreens, fenugreek is known as Samudra Methi in Maharashtra, especially in and around Mumbai, where it is often grown near the sea in the sandy tracts, hence the name (Samudra, "ocean" in Sanskrit). *Samudra Methi* is also grown in dry river beds in the Gangetic plains. When sold as a vegetable in India the young plants are harvested with their roots still attached and sold in small bundles in the markets and bazaars. Any remaining soil is washed off to extend their shelf life.

In Persian cuisine, fenugreek leaves are called shanbalile. They are the key ingredient and one of several greens incorporated into ghormeh sabzi and Eshkeneh, often said to be the Iranian national dishes.

Fenugreek is used in Eritrean and Ethiopian cuisine. The word for fenugreek in Amharic is *abesh* (or *abish*), and the seed is used in Ethiopia as a natural herbal medicine in the treatment of diabetes.

Palak (Spinach)

Spinach (*Spinacia oleracea*) is an edible flowering plant in the family of Amaranthaceae. It is native to central and southwestern Asia. It is an annual plant (rarely biennial), which grows to a height of up to 30 cm. Spinach may survive over winter in temperate regions. The leaves are alternate, simple, ovate to triangular-based, very variable in size from about 2–30 cm long and 1–15 cm broad, with larger leaves at the base of the plant and small leaves higher on the flowering stem. The flowers are inconspicuous, yellow-green, 3–4 mm diametre, maturing into a small, hard, dry, lumpy fruit cluster 5–10 mm across containing several seeds.

Common spinach, *Spinacia oleracea*, was long considered to be in the Chenopodiaceae family, but in 2003, the Chenopodiaceae family was combined with the Amaranthaceae family under the family name 'Amaranthaceae' in the order Caryophyllales. Within the Amaranthaceae family, Amaranthoideae and Chenopodioideae are now subfamilies, for the amaranths and the chenopods, respectively.

Spinach is thought to have originated in ancient Persia (modern Iran and neighbouring countries). Arab traders carried spinach into India, and then the plant was introduced into ancient China, where it was known as "Persian vegetable". The earliest available record of the spinach plant was recorded in Chinese, stating it was introduced into China via Nepal (probably in 647 AD).

In AD 827, the Saracens introduced spinach to Sicily. The first written evidence of spinach in the Mediterranean was recorded in three 10th-century works: the medical work by al-Razi (known as Rhazes in the West) and in two agricultural treatises, one by Ibn Wahshiya and the other by Qustus al-Rumi. Spinach became a popular vegetable in the Arab Mediterranean and arrived in Spain by the latter part of the 12th century, where the great Arab agronomist Ibn al-'Awwam called it the "captain of leafy greens". Spinach was also the subject of a special treatise in the 11th century by Ibn Hajjaj.

The prickly-seeded form of spinach was known in Germany by no later than the 13th century, though the smooth-seeded form was not described until 1552. (The smooth-seeded form is used in modern commercial production.)

Spinach first appeared in England and France in the 14th century, probably via Spain, and it gained quick popularity because it appeared in early spring, when other vegetables were scarce and when Lenten dietary restrictions discouraged consumption of other foods. Spinach is mentioned in the first known English cookbook, The *Forme of Cury* (1390), where it is referred to as *spinnedge* and/or *spynoches*. Smooth-seeded spinach was described in 1552.

In 1533, Catherine de' Medici became queen of France; she so fancied spinach, she insisted it be served at every meal. To this day, dishes made with spinach are known as "Florentine", reflecting Catherine's birth in Florence.

During World War I, wine fortified with spinach juice was given to French soldiers weakened by hemorrhage.

Culinary Information

Nutrition

Spinach has a high nutritional value and is extremely rich in antioxidants, especially when fresh, steamed, or quickly boiled. It is a rich source of vitamin A (and especially high in lutein), vitamin C, vitamin E, vitamin K, magnesium, manganese, folate, betaine, iron, vitamin B_2, calcium, potassium, vitamin B_6, folic acid, copper, protein, phosphorus, zinc, niacin, selenium and omega-3 fatty acids. Recently, opioid peptides called rubiscolins have also been found in spinach.

Polyglutamyl folate (vitamin B_9 or folic acid) is a vital constituent of cells, and spinach is a good source of folic acid. Boiling spinach can more than halve the level of folate left in the spinach, but microwaving does not affect folate content. Vitamin B_9 was first isolated from spinach in 1941.

Iron

Spinach, along with other green leafy vegetables, is considered to be rich in iron. For example, the United States Department of Agriculture states that a 180-g serving of boiled spinach contains 6.43 mg of iron, whereas a 170-g ground hamburger patty contains at most 4.42 mg. However, spinach contains iron absorption-inhibiting substances, including high levels of oxalate, which can bind to the iron to form ferrous oxalate and render much of the iron in spinach unusable by the body.

In addition to preventing absorption and use, high levels of oxalates remove iron from the body. But some studies have found that the addition of oxalic acid to the diet may improve iron absorption in rats over a diet with spinach without additional oxalic acid.

Calcium

Spinach also has a high calcium content. However, the oxalate content in spinach also binds with calcium, decreasing its absorption. Calcium and zinc also limit iron absorption. The calcium in spinach is the least bioavailable of calcium sources. By way of comparison, the human body can absorb about half of the calcium present in broccoli, yet only around 5% of the calcium in spinach.

Types of Spinach

A distinction can be made between older varieties of spinach and more modern ones. Older varieties tend to bolt too early in warm conditions. Newer varieties tend to grow more rapidly, but have less of an inclination to run up to seed. The older varieties have narrower leaves and tend to have a stronger and more bitter taste. Most newer varieties have broader leaves and round seeds.

The three basic types of spinach are:

- Savoy has dark green, crinkly and curly leaves. It is the type sold in fresh bunches in most supermarkets in the United States. One heirloom variety of savoy is *Bloomsdale*, which is somewhat resistant to bolting. Other common heirloom varieties are *Merlo Nero* (a mild variety from Italy) and *Viroflay* (a very large spinach with great yields).
- Flat- or smooth-leaf spinach has broad, smooth leaves that are easier to clean than Savoy. This type is often grown for canned

and frozen spinach, as well as soups, baby foods, and processed foods. *Giant Noble* is an example variety.

- Semi-savoy is a hybrid variety with slightly crinkled leaves. It has the same texture as Savoy, but it is not as difficult to clean. It is grown for both fresh market and processing. *Tyee Hybrid* is a common semi-savoy.

Production, Marketing and Storage

Spinach is sold loose, bunched, packaged fresh in bags, canned, or frozen. Fresh spinach loses much of its nutritional value with storage of more than a few days. While refrigeration slows this effect to about eight days, spinach will lose most of its folate and carotenoid content, so for longer storage, it is blanched and frozen, cooked and frozen, or canned. Storage in the freezer can be for up to eight months.

The Environmental Working Group reported spinach is one of the dozen most heavily pesticide-contaminated produce products. The most common pesticides found on spinach are permethrin, dimethoate, and DDT. Spinach is high in cadmium contamination. An FDA study found more in boiled spinach in the early 90's (0.125 mg/kg) than in the 320 other foods studied.

Spinach is packaged in air, or in nitrogen gas to extend shelf life. Some packaged spinach is exposed to radiation to kill any harmful bacteria that may be on the leaves. The Food and Drug Administration approves of irradiation of spinach leaves up to 4.0 kilograys; however, using radiation to sanitize spinach is of concern because it may deplete the leaves of their nutritional value. Researchers at the Agricultural Research Service experimentally tested the concentrations of vitamins C, E, K, B_9, and four other carotenoids in packaged spinach following irradiation. They found with increasing level of irradiation, four nutrients showed little or no change. Those nutrients include vitamins B_9, E, K, and the carotenoid neoxanthin. This study showed the irradiation of packaged spinach to have little or no change to the nutritional value of the crop, and the health benefits of irradiating packed spinach to reduce harmful bacteria seem to outweigh the loss of nutrients.

In Popular Culture

The cartoon character Popeye the Sailor Man is portrayed as having a strong affinity for spinach, becoming physically stronger after consuming it. The commonly accepted version of events states this portrayal was based on faulty calculations of the iron content. In the version, German scientist Emil von Wolff misplaced a decimal point in an 1870 measurement of spinach's iron content, leading to an iron value ten times higher than it should have been, and this faulty measurement was not noticed until the 1930s. This caused the popular misconception that spinach is high in iron that makes the body stronger.

Criminologist Mike Sutton wrote an article in the *Internet Journal of Criminology*, claiming the Popeye and iron link is just another long-standing myth, and spinach was chosen and promoted in Popeye for its vitamin A content alone. In the cited article, he also disputes the above — what he calls the *Spinach Popeye Iron Decimal Error Story (SPIDES)* — due to lack of verifiable sources, although he found a different reference from 1934 reporting twenty times the actual iron content. Further research by Sutton presents more evidence for the SPIDES being a myth which is cited as veracious evidence in the official errata pages for Samuel Arbesman's book, *The Half-life of Facts*.

Spinach, along with Brussels sprouts and other green vegetables, is often portrayed in the media, including children's shows, as being undesirable. For example, early Popeye movies played on the gag that Popeye's nephews never liked spinach but ate it anyway to grow up big and strong. The same theme is depicted in an even earlier classic cartoon from the New Yorker magazine, which also led to "I say it's spinach" becoming a 20th century American idiom for "nonsense".

Water Spinach

Ipomoea aquatica is a semiaquatic, tropical plant grown as a vegetable for its tender shoots and leaves. It is found throughout the tropical and subtropical regions of the world, although it is not known where it originated. This plant is known in English as water spinach, river spinach, water morning glory, water convolvulus, or by the more ambiguous names Chinese spinach, Chinese convolvulus, swamp cabbage or *kangkong* in Southeast Asia. Occasionally, it has also been mistakenly called "kale" in English, although kale is a strain of mustard belonging to the species *Brassica oleracea* and is completely unrelated to water spinach, which is a species of morning glory. It is known as *phak bung* in Thai, *rau muÑng* in Vietnamese, *trokuon* in Khmer, *kalmi shak* in Bengali, *kangkung* in Malay and Indonesian and *hayoyo* in Ghana.

I. aquatica grows in water or on moist soil. Its stems are 2–3 metres (7–10 ft) or more long, rooting at the nodes, and they are hollow and can float. The leaves vary from typically sagittate (arrow head-shaped) to lanceolate, 5–15 cm (2–6 in) long and 2–8 cm (0.8–3 in) broad. The flowers are trumpet-shaped, 3–5 cm (1–2 in) in diametre, and usually white in colour with a mauve centre. Propagation is either by planting cuttings of the stem shoots that will root along nodes or planting the seeds from flowers that produce seed pods.

Cultivation

Ipomoea aquatica is most commonly grown in East and Southeast Asia. It flourishes naturally in waterways and requires little, if any, care. It is used extensively in Burmese, Thai, Lao, Cambodian, Malay, Vietnamese, Filipino, and Chinese cuisine, especially in rural or *kampung* (village) areas. The vegetable is also extremely popular in Taiwan, where it grows well. During the Japanese occupation of Singapore in World War II, the vegetable grew remarkably easily in many areas, and became a popular wartime crop. In the Philippines, a variety of *kangkong* is grown in canals dug during the American occupation after the Spanish-American War, while another variety growing on land is called *Chinese kangkong*.

In non-tropical areas, it is easily grown in containers given enough water in a bright sunny location. It readily roots from cuttings.

Invasive Species

I. aquatica is listed on the USDA internet site as a "Class A noxious weed", especially in the states of Florida, California, and Hawaii, where it can be observed growing in the wild. *I. aquatica* has been extensively cultivated in Texas for over 30 years, having been originally brought there by Asian immigrants. Because no evidence indicates the plant has escaped into the wild, Texas lifted its ban on cultivation for personal use with no restrictions or requirements, noting its importance as a vegetable in many cultures, and also began permitting cultivation for commercial sales with the requirement of an exotic species permit. In Sri Lanka, it invades wetlands, where its long, floating stems form dense mats which can block the flow of water and prevent passage of boats.

Culinary Uses

The vegetable is a common ingredient in Southeast Asian dishes. In Singapore, Indonesia, and Malaysia, the tender shoots along with the leaves are usually stir-fried with chili pepper, garlic, ginger, dried shrimp paste (*belacan/terasi*) and other spices. In Penang and Ipoh, it is cooked with cuttlefish and a sweet and spicy sauce. Also known as *eng cai* in the Hokkien dialect, it can also be boiled with preserved cuttlefish, then rinsed and mixed with spicy *rojak* paste to become *jiu hu eng cai*. Boiled *eng cai* also can be served with fermented krill noodle *belacan bee hoon* and prawn noodle.

In the Philippines, where it is called *kangkóng*, the tender shoots are cut into segments and cooked, together with the leaves, in fish and meat stews, such as *sinigang*. The vegetable may also be eaten alone, such as in *adobong kangkóng*, where it is sautéed in cooking oil, onions, garlic, vinegar, soy sauce, and bouillon cube. A local appetiser called crispy *kangkóng* has the leaves coated in a flour-based batter and fried until crisp, similar to Japanese vegetable tempura.

Chinese cuisine has numerous ways of preparation, but a simple and quick stir-fry, either plain or with minced garlic, is probably the most common. In Cantonese, the water spinach is known as *ung3 coi3*, sometimes transliterated as *ong choy*. In Cantonese cuisine, a popular variation adds fermented bean curd. In Hakka cuisine, yellow bean paste is added, sometimes along with fried shallots.

In the cuisine of Cambodia, where it is called *trakuon*, water spinach is known to have a few types. The popular type is the usual local water plant, used in many traditional dishes. One type is known as Chinese *trakuon* grown as a straight stalk plant from soil. This type is known to be used in stir fry with pork or just with marinated soy beans (*sieng*). The Khmer popular dish *samlar machu* (sour soup) uses water spinach with fish or chicken. Water spinach is also eaten raw or parboiled along with other vegetables in dip dishes, e.g. *toeuk kroeung*, a sour/salty taste dish, with roughly minced fish mixed with lemon juice, crushed peanuts, basil, and *prahok* (fish paste) flavour, similar to *nam phrik*.

In Thailand, where it is called *phak bung*, it is eaten raw, often along with green papaya salad or *nam phrik*, in stir-fries and in curries such as *kaeng som*.

In Laos, where it is known as *pak bong*, and in Burma, where it is called *gazun ywet*, it is frequently stir-fried with oyster sauce or yellow soybean paste, and garlic and chillies.

In Vietnam, *I. aquatica* is a common ingredient and garnish in Vietnamese cuisine and was once served as a staple vegetable of the poor. In the South, the water spinach is julienned into thin strips and eaten with many kinds of noodles. It is also commonly cooked in a sour soup (*canh chua*), with tomatoes, other vegetables, and some kind of protein. *Rau muÑng* is also commonly sauteed with chopped garlic, oil (or pork fat), and fish sauce, known as *rau muÑng xào tÏi* (stir-fried water spinach with garlic) and served as a side dish in many meals.

In Bangladesh and West Bengal, it is known as *kolmishak* and stir-fried preparation of the leaves is a very popular dish.

Kale

Kale or borecole (*Brassica oleracea* Acephala Group) is a vegetable with green or purple leaves, in which the central leaves do not form a head. It is considered to be closer to wild cabbage than most domesticated forms.

The species *Brassica oleracea* contains a wide variety of vegetables, including broccoli, cauliflower, collard greens, and brussels sprouts. The cultivar group Acephala also includes spring greens and collard greens, which are extremely similar genetically. *Pieris rapae* is one of the most well known pests of the plant.

Cultivars

***Figure:** A bundle of curly-leaf kale.*

Kales can be classified by leaf type:

- Curly-leaved (Scots Kale)
- Plain-leaved
- Rape kale
- Leaf and spear (a cross between curly-leaved and plain-leaved kale)
- Cavolo nero (also known as black cabbage, Tuscan Cabbage, Tuscan Kale, Lacinato and dinosaur kale)

Because kale can grow well into winter, one variety of Rape Kale is called *Hungry Gap*, named after the period in winter in traditional agriculture when little else could be harvested.

An extra tall variety is known as Jersey kale or *cow cabbage*.

Culinary Uses

Kale freezes well and tastes sweeter and more flavourful after being exposed to a frost. Tender kale greens can provide an intense addition to salads, particularly when combined with other such strongly flavoured ingredients as dry-roasted peanuts, soy sauce-roasted almonds, red capsicum flakes, or a sesame-based dressing. When combined with oils or lemon juice, kale's flavour is noticeably reduced. When baked or dehydrated, kale takes on a consistency similar to that of a potato chip. Curly kale varieties are usually preferred for chips. The chips can be seasoned with salt or other spices.

Regional Culinary Uses

In the Netherlands, it is very frequently used in a traditional winter stamppot dish called boerenkool, which is a mix of kale and mashed potatoes, sometimes with fried bits of bacon added to it, and usually served with rookworst ("smoked sausage").

In Ireland, kale is mixed with mashed potatoes to make the traditional dish colcannon. It is popular on Halloween when it is sometimes served with sausages. Small coins are sometimes hidden inside as prizes.

In Italy, cavolo nero is an ingredient of the Tuscan soup ribollita.

A variety of kale, kai-lan, is a popular vegetable in China, Taiwan, and Vietnam, where it is commonly combined with beef dishes.

A traditional Portuguese soup, *caldo verde*, combines pureed potatoes, diced kale, olive oil, broth, and, generally, sliced cooked spicy sausage. In Brazil, where it was introduced by the Portuguese, it is an indispensable side dish for the national stew feijoada.

In the eastern African Great Lakes region, it is an essential ingredient in making a stew for ugali, which is almost always eaten with kale. Kale is also eaten throughout southeastern Africa, where it is typically boiled with coconut milk and ground peanuts and is served with rice or boiled cornmeal.

In Italy, kale (cavolo nero) is part of many dishes, such as "casseula" (pork stew), polenta (corn porridge) with kale, Parmesan cheese and olive oil and "pizzoccheri", buckwheat tagliatelle served with kale, melted fontina cheese and potatoes.

A whole culture around kale has developed in north-western Germany around the towns of Bremen, Oldenburg and Hannover. There, most social clubs of any kind will have a *Grünkohlessen or Kohlfahrt* ("kale tour") sometime between October and February, visiting a country inn to consume large quantities of kale stew, Pinkel sausage, Kassler, Mettwurst and Schnapps. These tours are often combined with a game of *Boßeln*. Most communities in the area have a yearly kale festival which includes naming a "kale king" (or queen).

Curly kale is used in Denmark and southwestern Sweden (Scania, Halland and Blekinge) to make *(grøn-)langkål* (Danish) or *långkål* (Swedish), an obligatory dish on the *julbord* in the region, and is commonly served together with the Christmas ham (Sweden). The kale is used to make a stew of minced boiled kale, stock, cream, pepper and salt that is simmered together slowly for a few hours. In Scotland, kale provided such a base for a traditional diet that the word in dialect Scots is synonymous with food. To be "off one's kail" is to feel too ill to eat. In Sweden, it is also commonly eaten as a soup, with a base of ham broth and the addition of onion and pork sausages.

In Montenegro, collards, kale, locally known as rashtan, is a favourite vegetable. It is particularly popular in the winter, cooked with smoked mutton (kastradina) and potatoes.

In the Southern United States, kale is often served braised, either alone or mixed with other greens, such as collard, mustard, or turnip. Flavoured kale chips have also been produced as a potato chip substitute.

In Japan, kale juice (known as *aojiru*) is a popular dietary supplement.

Collard Greens

Collard greens is the American English term for various loose-leafed cultivars of *Brassica oleracea*, part of the Acephala group which also contains cabbage and broccoli. The plants are grown for their large, dark-coloured, edible leaves and as a garden ornamental, mainly in Brazil, Portugal, the southern United States, many parts of Africa, Montenegro, Bosnia and Herzegovina, southern Croatia, northern Spain and in India. They are classified in the same cultivar group as kale and spring greens, to which they are genetically similar. The name "collard" is a corrupted form of the word "colewort" (the wild cabbage plant).

The plant is also called *couve* in Brazil and in Portugal, *couve galega* or *couve portuguesa* (among several other names) in Cape Verde, *berza* in Spanish-speaking countries, *raštika* in Bosnia and Herzegovina and Croatia and *raštan* in Montenegro and Serbia. In Kashmir, it is called *haak*. In Kenya it is more commonly known by its Swahili name, *sukuma wiki*, and is often confused with kales. In New Zealand, it is called Dalmatian Cabbage.

Description

The cultivar group name Acephala ("without a head" in Greek) refers to the fact that this variety of *B. oleracea* does not have the usual close-knit core of leaves (a "head") like cabbage. The plant is a biennial where winter frost occurs, and perennial in even colder regions. It is also moderately sensitive to salinity. It has an upright stalk, often growing up to two feet tall. The plant is very similar to kale. Popular cultivars of collard greens include 'Georgia Southern', 'Morris Heading', 'Butter Collard' (or *couve manteiga*), and *couve tronchuda.*

Figure: *Young collard plants growing in a container*

Cultivation and Storage

The plant is commercially cultivated for its thick, slightly bitter, edible leaves. They are available year-round, but are tastier and more nutritious in the cold months, after the first frost. For best texture, the leaves should be picked before they reach their maximum size, at which stage the leaves will be thicker and should be cooked differently from the new leaves. Age will not affect flavour. Flavour and texture also depend on the cultivar; the *couve manteiga* and *couve tronchuda* are especially appreciated in Brazil and Portugal.

Fresh collard leaves can be stored for up to 10 days if refrigerated to just above freezing (1°C) at high humidity (>95%). In domestic refrigerators, fresh collard leaves can be stored for about three days. Once cooked, they can be frozen and stored for greater lengths of time.

Nutritional Information

Widely considered to be a healthy food, collards are good sources of vitamin C and soluble fibre, and contain multiple nutrients with potent anticancer properties, such as diindolylmethane and sulforaphane. Roughly a quarter pound (approx. 100 g) of cooked collards contains 46 Calories. Collard greens are also a high source of vitamin K (the clotting vitamin) and should be eaten in moderation by individuals taking blood thinners.

Researchers at the University of California at Berkeley have recently discovered that 3,3'-diindolylmethane in *Brassica* vegetables such as collard greens is a potent modulator of the innate immune response system with potent antiviral, antibacterial and anticancer activity.

Culinary Use

Collard greens have been eaten for at least 2000 years, with evidence showing that the Ancient Greeks cultivated several forms of both collard greens and kale.

Southern United States

Collard greens are a staple vegetable of Southern U.S. cuisine. They are often prepared with other similar green leaf vegetables, such as kale, turnip greens, spinach, and mustard greens in "mixed greens". They are generally eaten year-round in the South.

Typical seasonings when cooking collards can consist of smoked and salted meats (ham hocks, smoked turkey drumsticks, pork neckbones, fatback or other fatty meat), diced onions, vinegar, salt, and black, white, or crushed red pepper, and some cooks add a small amount of sugar.

Traditionally, collards are eaten on New Year's Day, along with black-eyed peas or field peas and cornbread, to ensure wealth in the coming year, as the leaves resemble folding money. Cornbread is used to soak up the "pot liquor", a nutrient-rich collard broth. Collard greens may also be thinly sliced and fermented to make collard kraut, which is often cooked with flat dumplings.

Kenya (East Africa)

Collard greens are known as 'sukuma wiki' in Kenya.

In Congo, Tanzania and Kenya (East Africa), thinly sliced collard greens are the main accompaniments of a popular dish known as sima or ugali (a corn flour cake).

'Sukuma Wiki' is mainly lightly sauteed in oil until tender, flavoured in onions and seasoned with salt and is served either as the main accompaniment or as a side dish with preferred meat (fish, chicken, beef, pork).

Brazil and Portugal

In Portuguese and Brazilian cuisine, collard greens (or *couve*) are common accompaniments of fish and meat dishes. They are a standard side dish for *feijoada*, a popular pork and beans-style stew.

Thinly sliced collard greens are also the main ingredient of a popular Portuguese soup, *caldo verde* (green broth). For this broth the leaves are sliced into strips, 2 or 3 mm wide (sometimes by the grocer or market vendor, with a special hand-cranked slicer) and added to the other ingredients 15 minutes before it is served.

India

In Kashmir this vegetable is well renowned. It finds its place in almost every meal in Kashmiri houses, with both leaves and roots consumed. Leaves in the bud are harvested by pinching in early spring, when the dormant buds sprout and give out tender leaves. Also, seedlings of 35–40 days' age, as well as mature plants, are pulled out along with roots from thickly sown beds. When the extending stem bears alternate leaves in quick succession during on-season, older leaves are harvested periodically. Before the autumn season, the apical portion of stem is removed along with the whorled leaves.It is called *haak* there.

The roots and the leaves may be cooked together or separately. A common dish is *haak rus*, a soup of whole collard leaves cooked in water, salt and oil, usually consumed with rice. The leaves are also cooked along with meat, fish or cheese. In the winter, collard leaves and roots are fermented to form a very popular pickle called *haak-e-aanchaar*.

Sorrel

Growing sorrel yields great benefits, in part because the cold-hardy plants return year after year. Both garden sorrel and French sorrel have a unique lemony flavour much prized in spring salads and sorrel soup. This guide includes descriptions of the types of sorrel with tips for growing sorrel in your garden.

Common sorrel or garden sorrel (*Rumex acetosa*), often simply called sorrel, is a perennial herb in the family Polygonaceae. Other names for sorrel include spinach dock and narrow-leaved dock. It is a common plant in grassland habitats and is cultivated as a garden herb or leaf vegetable (pot herb).

Growth

***Figure:** Sorrel (*aveluk *in Armenian) picked fresh from Mount Ara and braided before selling for ease of drying and extended use.*

Sorrel is a slender plant about 60 cm high, with roots that run deep into the ground, as well as juicy stems and edible, oblong leaves. The lower leaves are 7 to 15 cm in length, slightly arrow-shaped at the base, with very long petioles. The upper ones are sessile, and frequently become crimson. The leaves are eaten by the larvae of several species of Lepidóptera (butterfly and moth) including the blood-vein moth.

Figure: *Sorrel soup with egg and croutons; Polish cuisine.*

Characteristics

It has whorled spikes of reddish-green flowers, which bloom in summer, becoming purplish. The stamens and pistils are on different plants (dioecious); the ripe seeds are brown and shining.

Distribution

Rumex acetosa occurs in grassland habitats throughout Europe from the northern Mediterranean coast to the north of Scandinavia and in parts of Central Asia.

Uses

Common sorrel has been cultivated for centuries. The leaves may be puréed in soups and sauces or added to salads; they have a flavour that is similar to kiwifruit or sour wild strawberries. The plant's sharp taste is due to oxalic acid. In small quantities sorrel is harmless; in large quantities it can be fatal.

In northern Nigeria, sorrel is known as *yakuwa* or *sure* (pronounced suuray) in Hausa or *karassu* in Kanuri. It is also used in stews usually in addition to spinach. In some Hausa communities, it is steamed and made into salad using *kuli-kuli* (traditional roasted peanut cakes with oil extracted), salt, pepper, onion and tomatoes. The recipe varies according to different levels of household income. A drink called *solo* is made from a decoction of the plant calyx.

In Romania, wild or garden sorrel, known as *măcri°* or *°tevie,* is used to make sour soups, stewed with spinach, added fresh to lettuce and spinach in salads or over open sandwiches.

In Russia and Ukraine it is called shchavel' (ùàâåëü) and is used to make soup called green borscht. It is used as a soup ingredient in other countries, too (e.g., Lithuania, where it is known as rûgðtynë).

In Hungary the plant and its leaves is known as *sóska*. It is called *kuzukulaðý* ('lamb's ear') in Turkish. In Polish it is called *szczaw*.

In Croatia and Bulgaria is used for soups or with mashed potatoes, or as part of a traditional dish containing eel and other green herbs.

In rural Greece it is used with spinach, leeks, and chard in spanakopita.

In the Flemish speaking part of Belgium it is called "zurkel" and canned pureed sorrel is mixed with mashed potatoes and eaten with sausages, meatballs or fried bacon, as a traditional winter dish.

In Vietnam it is called Rau Chua and is used to added fresh to lettuce and in salads for Bánh Xço.

In Portugal, it's called "azeda" (sour), and is usually chewed raw.

In India, the leaves are called chukkakura in Telugu and are used in making delicious recipes.Chukkakura pappu soup made with yellow lentils [Also called toor dal in India]

In Albania it is called *lëpjeta*, the leaves are simmered and served cold marinated in olive oil, it is used in soups, and even as an ingredient for filling byrek pies (*byrek me lakra*).

Leeks

Almost any time of year is good for growing leeks, the most upright members of the onion family. Summer leeks are fast to grow from seed started indoors in late winter, or you can start leek seedlings in late summer to grow from fall to spring.

The leek is a vegetable that belongs, along with onion and garlic, to the genus *Allium*, currently placed in family Amaryllidaceae, subfamily Allioideae.

Historically many scientific names were used for leeks, which are now treated as cultivars of *Allium ampeloprasum*. Two related vegetables, elephant garlic and Kurrat, are also cultivars of *A. ampeloprasum*, although different in their uses as food.

The edible part of the leek plant is a bundle of leaf sheaths that is sometimes erroneously called a *stem* or *stalk*.

Rather than forming a tight bulb like the onion, the leek produces a long cylinder of bundled leaf sheaths that are generally blanched by pushing soil around them (trenching). They are often sold as small seedlings in flats that are started off early in greenhouses, to be planted out as weather permits. Once established in the garden, leeks are hardy; many varieties can be left in the ground during the winter to be harvested as needed.

Cultivars

Leek cultivars may be treated as a single cultivar group, e.g. as *Allium ampeloprasum* 'Leek Group'. The cultivars can be subdivided in several ways, but the most common types are "summer leeks", intended for harvest in the season when planted, and overwintering leeks, meant to be harvested in the spring of the year following planting. Summer leek types are generally smaller than overwintering types; overwintering types are generally more strongly flavoured. Cultivars include 'King Richard' and 'Tadorna Blue'.

Growing

Leeks are easy to grow from seed and tolerate standing in the field for an extended harvest. Leeks usually reach maturity in the autumn months, and they have few pest or disease problems. Leeks can be bunched and harvested early when they are about the size of a finger or pencil, or they can be thinned and allowed to grow to a much larger mature size. Hilling leeks can produce better specimens.

Cuisine

Leeks have a mild onion-like taste. In its raw state, the vegetable is crunchy and firm. The edible portions of the leek are the white base of the leaves (above the roots and stem base), the light green parts, and to a lesser extent the dark green parts of the leaves. One of the most popular uses is for adding flavour to stock.

The dark green portion is usually discarded because it has a tough texture, but it can be sauteed or added to stock. A few leaves are sometimes tied with twine and other herbs to form a *bouquet garni*.

Leeks are typically chopped into slices 5–10 mm thick. The slices have a tendency to fall apart, due to the layered structure of the leek. The different ways of preparing the vegetable are:

- Boiled, which turns it soft and mild in taste. (Care should be taken to chop the vegetable, or else the intact fibres that run the length of the vegetable will tangle into a ball while chewing.)
- Fried, which leaves it crunchier and preserves the taste.
- Raw, which can be used in salads, doing especially well when they are the prime ingredient.
- In Turkish cuisine, leeks are chopped into thick slices, then boiled and separated into leaves and finally filled with a filling usually containing rice, herbs (generally parsley and dill), onion and black pepper. For sarma with olive oil, currants, pine nuts and cinnamon are added and for sarma with meat, minced meat is added to the filling.

  In Turkey, especially "Zeytinyaðlý pýrasa" (leek with olive oil), "Ekþili pýrasa (sour leek), "Etli pýrasa" (leek with meat), "Pýrasa Musakka"(leek musakka), "Pýrasalý börek (börek with leek) and "Pýrasa köftesi" leek meatball are also cooked.

Leeks are an ingredient of cock-a-leekie soup, leek and potato soup, and *vichyssoise*, as well as plain leek soup.

Because of their symbolism in Wales, they have come to be used extensively in that country's cuisine. Elsewhere in Britain, leeks have come back into favour only in the last 50 years or so, having been overlooked for several centuries.

Historical Consumption

Bible commentators attribute the çöéø specimen - acclaimed by the Israelites to be of abundance in Egypt - as the leek Dried specimens from archaeological sites in ancient Egypt, as well as wall carvings and drawings, led Zohary and Hopf to conclude the leek was a part of the Egyptian diet from at least the second millennium BCE onwards. They also allude to surviving texts that show it had been also grown in Mesopotamia from the beginning of the second millennium BCE. The leek was the favourite vegetable of the Emperor Nero, who consumed it in soup or in oil, believing it beneficial to the quality of his voice.

Cultural Significance

The leek is one of the national emblems of Wales, worn along with the daffodil (in Welsh, the daffodil is known as "Peter's leek," *Cenhinen Bedr*) on St. David's Day. According to one legend, King Cadwaladr of Gwynedd ordered his soldiers to identify themselves by wearing the vegetable on their helmets in an ancient battle against the Saxons that took place in a leek field. This story may have been made up by the English poet Michael Drayton, but the leek has been known to be a symbol of Wales for a long time; Shakespeare, for example, refers to the custom of wearing a leek as an "ancient tradition" in *Henry V*. In the play, Henry tells Fluellen that he is wearing a leek "for I am Welsh, you know, good countryman." The 1985 and 1990 British one pound coins bear the design of a leek in a coronet, representing Wales.

Alongside the other national floral emblems of countries in the Commonwealth (including the English Tudor Rose, Scottish thistle, Irish shamrock, Canadian maple leaf, and Indian lotus), the Welsh leek appeared on the coronation gown of Elizabeth II. It was designed by Norman Hartnell; when Hartnell asked if he could exchange the leek for the more aesthetically pleasing Welsh daffodil, he was told no.

Perhaps the most visible use of the leek, however, is as the cap badge of the Welsh Guards, a regiment within the Household Division of the British Army.

In Romania, the leek is also widely considered a symbol of Oltenia, a historical region in the southwestern part of the country.

Lettuce

Lettuce (*Lactuca sativa*) is an annual plant of the aster or sunflower family Asteraceae. It is most often grown as a leaf vegetable, but sometimes for its stem and seeds. Lettuce was first cultivated by the ancient Egyptians who turned it from a weed, whose seeds were used to produce oil, into a plant grown for its leaves. Lettuce spread to the Greeks and Romans, the latter of whom gave it the name "*lactuca*", from which the English "lettuce" is ultimately derived. By 50 AD, multiple types were described, and lettuce appeared often in medieval writings, including several herbals. The 16th through 18th centuries saw the development of many varieties in Europe, and by the mid-18th century cultivars were described that can still be found in gardens. Europe and North America originally dominated the market for lettuce, but by the late 1900s the consumption of lettuce had spread throughout the world.

Generally grown as a hardy annual, lettuce is easily cultivated, although it requires relatively low temperatures to prevent it from flowering quickly. It can be plagued with numerous nutrient deficiencies, as well as insect and mammal pests and fungal and bacterial diseases. *L. sativa* crosses easily within the species and with some other species within the *Lactuca* genus; although this trait can be a problem to home gardeners who attempt to save seeds, biologists have used it to broaden the gene pool of cultivated lettuce varieties. World production of lettuce and chicory for calendar year 2010 stood at 23,620,000 metric tons (23,250,000 long tons; 26,040,000 short tons), over half of which came from China.

Lettuce is most often used for salads, although it is also seen in other kinds of food, such as soups, sandwiches and wraps; it can also be grilled. One variety, the Woju (′*f*ã,) or asparagus lettuce, is grown for its stems, which are eaten either raw or cooked. Lettuce is a good source of vitamin A and potassium, as well as a minor source for several other vitamins and nutrients. Despite its beneficial properties, contaminated lettuce is often a source of bacterial, viral and parasitic outbreaks in humans, including *E. coli* and *Salmonella*. In addition to its main use as a leafy green, it has also gathered religious and medicinal significance over centuries of human consumption.

Description

Lettuce's native range spreads from the Mediterranean to Siberia, although it has been transported to almost all areas of the world. Plants generally have a height and spread of 6 to 12 inches (15 to 30 cm). The leaves are colourful, mainly in the green and red colour spectrums, with some variegated varieties. There are also a few varieties with yellow, gold or blue-teal leaves. Lettuces have a wide range of shapes and textures, from the dense heads of the iceberg type to the notched, scalloped, frilly or ruffly leaves of leaf varieties. Lettuce plants have a root system that includes a main taproot and smaller secondary roots.

Some varieties, especially those found in the US and Western Europe, have long, narrow taproots and a small set of secondary roots. Longer taproots and more extensive secondary systems are found in varieties from Asia.

Figure: *Mature lettuce inflorescence in fruit*

Depending on the variety and time of year, lettuce generally lives 65–130 days from planting to harvesting. Because lettuce that flowers (through the process known as "bolting") becomes bitter and unsaleable, plants grown for consumption are rarely allowed to grow to maturity. Lettuce flowers more quickly in hot temperatures, while freezing temperatures cause slower growth and sometimes damage to outer leaves. Once plants move past the edible stage, they develop flower stalks up to 3 feet (0.9 m) high with small yellow blossoms. Like other members of the tribe Cichorieae, lettuce inflorescences (also known as flower heads or capitula) are composed of multiple florets, each with a modified calyx called a pappus (which becomes the feathery "parachute" of the fruit), a corolla of five petals fused into a ligule or strap, and the reproductive parts.

These include fused anthers that form a tube which surrounds a style and bipartite stigma. As the anthers shed pollen, the style elongates to allow the stigmas, now coated with pollen, to emerge from the tube. The ovaries form compressed, obovate (teardrop-shaped) dry fruits that do not open at maturity, measuring 3 to 4 mm long. The fruits have 5–7 ribs on each side and are tipped by two rows of small white hairs.

The pappus remains at the top of each fruit as a dispersal structure. Each fruit contains one seed, which can be white, yellow, gray or brown depending on the variety of lettuce.

The domestication of lettuce over the centuries has resulted in several changes through selective breeding: delayed bolting, larger seeds, larger leaves and heads, better taste and texture, a lower latex content, and different leaf shapes and colours. Work in these areas continues through the present day. Scientific research into the genetic modification of lettuce is ongoing, with over 85 field trials taking place between 1992 and 2005 in the European Union and United States to test modifications allowing greater herbicide tolerance, greater resistance to insects and fungi and slower bolting patterns. However, genetically modified lettuce is not currently used in commercial agriculture.

History

Figure: *Romaine lettuce, a descendant of some of the earliest cultivated lettuce*

Lettuce was first cultivated in ancient Egypt for the production of oil from its seeds. This plant was probably selectively bred by the Egyptians into a plant grown for its edible leaves, with evidence of its cultivation appearing as early as 2680 BC. Lettuce was considered a sacred plant of the reproduction god Min, and it was carried during his festivals and placed near his images. The plant was thought to help the god "perform the sexual act untiringly." Its use in religious ceremonies resulted in the creation of many images in tombs and wall paintings. The cultivated variety appears to have been about 30 inches (76 cm) tall and resembled a large version of the modern romaine lettuce. These upright lettuces were developed by the Egyptians and passed to the Greeks, who in turn shared them with the Romans. Circa 50 AD, Roman agriculturalist Columella described several lettuce varieties – some of which may have been ancestors of today's lettuces.

Lettuce appears in many medieval writings, especially as a medicinal herb. Hildegard of Bingen mentioned it in her writings on medicinal herbs between 1098 and 1179, and many early herbals also describe its uses. In 1586, Joachim Camerarius provided descriptions of the three basic modern lettuces – head lettuce, loose-leaf lettuce and romaine or cos lettuce. Lettuce was first brought to the Americas from Europe by Christopher Columbus in the late 15th century. Between the late 16th century and the early 18th century, many varieties were developed in Europe, particularly Holland. Books published in the mid-18th and early 19th centuries describe several varieties found in gardens today.

Due to its short life span after harvest, lettuce was originally sold relatively close to where it was grown. The early 1900s saw the development of new packing, storage and shipping technologies that improved the lifespan and transportability of lettuce and resulted in a significant increase in availability. During the 1950s, lettuce production was revolutionized with the development of vacuum cooling, which allowed field cooling and packing of lettuce, replacing the previously used method of ice-cooling in packing houses outside the fields.

Lettuce is very easy to grow, and as such has been a significant source of sales for many seed companies. Tracing the history of many varieties is complicated by the practice of many companies, particularly in the US, of changing a variety's name from year to year. This was done for several reasons, the most prominent being to boost sales by promoting a "new" variety or to prevent customers from knowing that the variety had been developed by a competing seed company. Documentation from the late 19th century shows between 65 and 140 distinct varieties of lettuce, depending on the amount of variation allowed between types – a distinct difference from the 1,100 named lettuce varieties on the market at the time. Names also often changed significantly from country to country. Although most lettuce grown today is used as a vegetable, a minor amount is used in the production of tobacco-free cigarettes; however, domestic lettuce's wild relatives produce a leaf that visually more closely resembles tobacco.

Cultivation

A hardy annual, some varieties of lettuce can be overwintered even in relatively cold climates under a layer of straw, and older, heirloom varieties are often grown in cold frames. Lettuces meant for the cutting of individual leaves are generally planted straight into the garden in thick rows. Heading varieties of lettuces are commonly started in flats, then transplanted to individual spots, usually 8 to 14 inches (20 to 36 cm) apart, in the garden after developing several leaves. Lettuce spaced further apart receives more sunlight, which improves colour and nutrient quantities in the leaves. Pale to white lettuce, such as the centres in some iceberg lettuce, contain few nutrients.

Figure: *A transplanted bed of lettuce in a greenhouse*

Lettuce grows best in full sun in loose, nitrogen-rich soils with a pH of between 6.0 and 6.8. Heat generally prompts lettuce to bolt, with most varieties growing poorly above 75 °F (24 °C); cool temperatures prompt better performance, with 60 to 65 °F (16 to 18 °C) being preferred and as low as 45 °F (7 °C) being tolerated. Plants in hot areas that are provided partial shade during the hottest part of the day will bolt more slowly. Temperatures above 80 °F (27 °C) will generally result in poor or non-existent germination of lettuce seeds. After harvest, lettuce lasts the longest when kept at 32 °F (0 °C) and 96 percent humidity. Lettuce quickly degrades when stored with fruit such as apples, pears and bananas that release the ripening agent ethylene gas. The high water content of lettuce (94.9 percent) creates problems when attempting to preserve the plant – it cannot be successfully frozen, canned or dried and must be eaten fresh.

Lettuce varieties will cross with each other, making spacing of 5 to 20 feet (1.5 to 6.1 m) between varieties necessary to prevent contamination when saving seeds. Lettuce will also cross with *Lactuca serriola* (wild lettuce), with the resulting seeds often producing a plant with tough, bitter leaves. Celtuce, a lettuce variety grown primarily in Asia for its stems, crosses easily with lettuces grown for their leaves. This propensity for crossing, however, has led to breeding programmes using closely related species in *Lactuca*, such as *L. serriola*, *L. saligna*, and *L. virosa*, to broaden the available gene pool. Starting in the 1990s, such programmes began to include more distantly related species such as *L. tatarica*. Seeds keep best when stored in cool conditions, and, unless stored cryogenically, remain viable the longest when stored at –4 °F (–20 °C); they are relatively short lived in storage.

At room temperature, lettuce seeds remain viable for only a few months. However, when newly harvested lettuce seed is stored cryogenically, this life increases to a half-life of 500 years for vaporized nitrogen and 3,400 years for liquid nitrogen; this advantage is lost if seeds are not frozen promptly after harvesting.

Cultivars

There are several types of lettuce, but three (leaf, head and cos or romaine) are the most common. There are seven main cultivar groups of lettuce, each including many varieties:

- Leaf – Also known as looseleaf, cutting or bunching lettuce, this type has loosely bunched leaves and is the most widely planted. It is used mainly for salads.
- Romaine/Cos – Used mainly for salads and sandwiches, this type forms long, upright heads. This is the most often used lettuce in Caesar salads.
- Crisphead – Better known as "iceberg" lettuce, the most popular lettuce in the US, this type is very heat-sensitive and was originally adapted for growth in the northern US. It ships well, but is low in flavour and nutritional content, being composed of even more water than other lettuce types.
- Butterhead – Also known as Boston or Bibb lettuce, this type is a head lettuce with a loose arrangement of leaves, known for its sweet flavour and tender texture.
- Summercrisp – Also called Batavian or French Crisp, this lettuce is midway between the crisphead and leaf types. These lettuces tend to be larger, bolt-resistant and well-flavoured.
- Stem – This type is grown for its seedstalk, rather than its leaves, and is used in Asian cooking, primarily Chinese, as well as stewed and creamed dishes.

- Oilseed – This type is grown for its seeds, which are pressed to extract an oil mainly used for cooking. It has few leaves, bolts quickly and produces seeds around 50 percent larger than other types of lettuce.

The butterhead and crisphead types are sometimes known together as "cabbage" lettuce, because their heads are shorter, flatter, and more cabbage-like than romaine lettuces.

Cultivation Problems

Nutrient deficiencies, including a lack of boron, phosphorus, calcium, lithium, molybdenum or copper, can cause a variety of plant problems that range from malformed plants to a lack of head growth. Many insects are attracted to lettuce, including cutworms, which cut seedlings off at the soil line; wireworms and nematodes, which cause yellow, stunted plants; tarnished plant bugs and aphids, which cause yellow, distorted leaves; leafhoppers, which cause stunted growth and pale leaves; thrips, which turn leaves gray-green or silver; leafminers, which create tunnels within the leaves; flea beetles, which cut small holes in leaves and caterpillars, slugs and snails, which cut large holes in leaves. Mammals, including rabbits and groundhogs, also eat the plants. Lettuce contains several defensive compounds, including sesquiterpene lactones, and other natural phenolics such as flavonol and glycosides, which help to protect it against pests. Certain varieties contain more than others, and some selective breeding and genetic modification studies have focused on using this trait to identify and produce commercial varieties with increased pest resistance.

Lettuce also suffers from several viral diseases, including big vein, which causes yellow, distorted leaves, and mosaic virus, which is spread by aphids and causes stunted plant growth and deformed leaves. Aster yellows are a disease-causing bacteria carried by leafhoppers, which causes deformed leaves.

Fungal diseases include powdery mildew and downy mildew, which cause leaves to mold and die and bottom rot, lettuce drop and gray mold, which cause entire plants to rot and collapse. Crowding lettuce tends to attract pests and diseases. Weeds can also be an issue, as cultivated lettuce is generally not competitive with them, especially when directly seeded into the ground. Transplanted lettuce (started in flats and later moved to growing beds) is generally more competitive initially, but can still be crowded later in the season, causing misshapen lettuce and lower yields. Weeds also act as homes for insects and disease and can make harvesting more difficult. Herbicides are often used to control weeds in commercial production. However, this has led to the development of herbicide-resistant weeds and prompted environmental and health concerns.

Production

Lettuce is the only member of the *Lactuca* genus to be grown commercially. The Food and Agriculture Organization of the United Nations (FAO) reports that world production of lettuce and chicory (the two crops are combined by the FAO for reporting purposes) for calendar year 2010 was 23,622,366 metric tons (23,249,287 long tons; 26,039,201 short tons). This came primarily from China (53 percent), the US (17 percent) and India (4 percent). Although China is the top world producer of lettuce, the majority of the crop is consumed domestically. Spain is the world's largest exporter of lettuce, with the US ranking second.

Western Europe and North America were the original major markets for large-scale lettuce production. By the late 1900s, Asia, South America, Australia and Africa became more substantial markets. Different locations tended to prefer different types of lettuce, with butterhead prevailing in northern Europe and Great Britain, romaine in the Mediterranean and stem lettuce in China and Egypt. By the late 20th century, the preferred types began to change, with crisphead, especially iceberg, lettuce becoming the dominant type in northern Europe and Great Britain and more popular in western Europe. In the US, no one type predominated until the early 20th century, when crisphead lettuces began gaining popularity. After the 1940s, with the development of iceberg lettuce, 95 percent of the lettuce grown and consumed in the US was crisphead lettuce. By the end of the century, other types began to regain popularity and eventually made up over 30 percent of production. Stem lettuce was first developed in China, and remains primarily cultivated in that country.

Lettuce production methods, including all of the processes from growing to sales, have become much larger in scale during the 20th century. The majority of agricultural production is done with the application of large amounts of chemicals, including fertilizers and pesticides, but organic production makes up a growing percentage of the market – a trend that began with small growers but moved to a more industrial scale. More non-heading types, mostly leaf and romaine lettuces, are also being grown. In the first years of the 21st century, bagged salad products began to hold a growing portion of the lettuce market, especially in the US. Processed from what was previously waste lettuce not considered acceptable for the fresh market, these products are packaged in a manner that makes them last longer than standard head lettuce after harvest. As of 2007, 70 percent of the lettuce production in the US came from California; in that country it ranks third in produce consumption behind tomatoes and oranges.

Culinary Use

As described around 50 AD, lettuce leaves were often cooked and served by the Romans with an oil-and-vinegar dressing; however, smaller leaves were sometimes eaten raw. During the 81–96 AD reign of Domitian, the tradition of serving a lettuce salad before a meal began. Post-Roman Europe continued the tradition of poaching lettuce, mainly with large romaine types, as well as the method of pouring a hot oil and vinegar mixture over the leaves. Today, the majority of lettuce is grown for its leaves, although one type is grown for its stem and one for its seeds, which are made into an oil. Most lettuce is used in salads, either alone or with other greens, vegetables, meats and cheeses. Romaine lettuce is often used for Caesar salads, with a dressing that includes anchovies and eggs. Lettuce leaves can also be found in soups, sandwiches and wraps, while the stems are eaten both raw and cooked. The consumption of lettuce in China developed differently from in Western countries, due to health risks and cultural aversion to eating raw leaves. In that country, "salads" were created from cooked vegetables and served hot or cold. Lettuce was also used in a larger variety of dishes than in Western countries, contributing to a range of dishes including bean curd and meat dishes, soups and stir-frys plain or with other vegetables. Stem lettuce, widely consumed in China, is eaten either raw or cooked, the latter primarily in soups and stir-frys.

Nutrition and Health

Depending on the variety, lettuce is a good source of vitamin A, vitamin K and potassium, with higher concentrations of vitamin A found in darker green lettuces. It also provides some dietary fibre (concentrated in the spine and ribs), carbohydrates, protein and a small amount of fat. With the exception of the iceberg type, lettuce also provides some vitamin C, calcium, iron and copper, with vitamins and minerals largely found in the leaf. Lettuce naturally absorbs and concentrates lithium.

Food-borne Illness

Although most food-borne pathogens can survive on stored lettuce, they tend to decline in number during the storage period. The exception to this is *Listeria monocytogenes*, the causative agent of listeriosis, which multiplies in storage. However, despite very high levels of the bacteria being found on ready-to-eat lettuce products, a 2008 study found no incidences of food-borne illness related to listeriosis. The researcher posited that this may be due to the product's short shelf life, indigenous microflora competing with the *Listeria* bacteria or possible properties within the lettuce that cause the bacteria to be unable to cause listeriosis.

Religious and Medicinal Lore

In addition to its usual purpose as an edible leafy vegetable, lettuce has had a number of uses in ancient (and even some more modern) times as a medicinal herb and religious symbol. For example, ancient Egyptians thought lettuce to be a symbol of sexual prowess and a promoter of love and childbearing in women. The Romans likewise claimed that it increased sexual potency. In contrast, the ancient Greeks connected the plant with male impotency, and served it during funerals (probably due to its role in the myth of Adonis's death), and British women in the 1800s believed it would cause infertility and sterility. Lettuce has mild narcotic properties – it was called "sleepwort" by the Anglo-Saxons because of this attribute – although the cultivated *L. sativa* has lower levels of the narcotic than its wild cousins. This narcotic effect is a property of two sesquiterpene lactones which are found in the white liquid (latex) in the stems of lettuce, called lactucarium or "lettuce opium".

Lettuce is also eaten as part of the Jewish Passover Seder, where it is considered the optimal choice for use as the bitter herb which is eaten together with the matzah.

Lettuce extracts are sometimes used in skin creams and lotions for treating sunburn and rough skin. It was once thought to be useful in relieving liver issues. Some American settlers claimed that smallpox could be prevented through the ingestion of lettuce, and an Iranian belief suggested consumption of the seeds when afflicted with typhoid. Folk medicine has also claimed it as a treatment for pain, rheumatism, tension and nervousness, coughs and insanity; scientific evidence of these benefits in humans has not been found, although some similar effects have been demonstrated in mice and toads. The religious ties of lettuce continue into the present day among the Yazidi people of northern Iraq, who have a religious prohibition against eating the plant.

Pest Control in Leafy Vegetables

Leafminers

Plants are often disfigured and damaged by several species of small flies that live as maggots between the upper and lower surfaces of the leaves. Their feeding causes large white blotches and winding trails through the interior of the leaf. Infected leaves are unattractive and unfit for human consumption.

The preferred hosts of the spinach leafminer are spinach, beet, and chard. The insect also attacks many species of weeds. Adult flies emerge in April and deposit eggs on the underside of the host plant's leaves. The eggs hatch in four to six days, and the young maggots bore directly into the leaves where they feed for 10 to 14 days. There are three generations each year.

Control : Eliminating weeds will aid in the control of leafminers. During most years sprays are needed to prevent injury. Consult the most recent Commercial Vegetable Production Guide (available at your County Extension Office) for specific control measures. Follow directions on the labels according to plant type when applying insecticides.

Aphids

Aphids are small, soft-bodied insects, often called plant lice. They spread several virus diseases, reduce vigor and yield of plants, and contaminate leaves. Sometimes natural controls, such as beneficial parasites, hold down aphid population, especially if you have not recently used a broad-spectrum pesticide.

Control: Eliminating weeds will aid in the control of aphids. Consult the most recent Commercial Vegetable Production Guide for specific control measures. Follow directions on the labels according to plant type when applying insecticides.

Cabbage Worms

Two worms attack leafy vegetables: the cabbage looper and the imported cabbage worm . The tiny, light green worms are called loopers because of their characteristic way of walking. The looping movement results from having only two pairs of legs towards the tail end of the body. Loopers do not overwinter in Pennsylvania, so problems vary from year to year.

However, it is during late August and September that the looper can cause considerable injury. The imported cabbage worm is velvety green with numerous ridges across the body. The worms have four pairs of legs on the centre of the body. The cabbage worm is a persistent problem, from early spring until frost. The adult insect is the common white butterfly often seen flying around cabbage plants.

Control: Avoid spraying directly over the tops of plants since most eggs and young loopers feed on the underside of leaves. Pest control programmes for home garden vegetables use both cultural and chemical control measures. The success or failure of a fungicide or insecticide is related to correctly identifying the pest problem, the method of application, weather conditions, correct timing of sprays, and choosing the right pesticide. Follow directions on the labels according to plant type when applying insecticides.

Warning

Pesticides are poisonous. Read and follow directions and safety precautions on labels. Handle carefully and store in original labelled containers out of the reach of children, pets, and livestock. Dispose of empty containers right away, in a safe manner and place. Do not contaminate forage, streams, or ponds.

6

Root Vegetables

Root vegetables are underground plant parts used as vegetables. They are called root vegetables for lack of a better generic term, but include both true roots such as tuberous roots and taproots, as well as non-roots such as tubers, rhizomes, corms, bulbs, and hypocotyls.

Potatoes

The world's favourite root vegetable, the potato comes in innumerable varieties. A member of the nightshade family, like tomatoes and aubergines, it originated in South America and has been grown in Europe since the 16th century.

Shapes vary from small ('finger') potatoes like Anya to large, round types like the King Edward. Most have pale brown skins and cream or yellow flesh, but some speciality varieties are differently coloured, like the Purple Peruvian.

'Waxy' potatoes such as Charlotte are great used in salads, while 'floury' potatoes such as Maris Piper are ideal for mash and baking.

Native to the mountains of South America, potatoes should be planted first thing in spring, when the soil is still cool. Gardeners can tap into a deliciously diverse selection of varieties, and it's easy to save and replant your favourite varieties from one year to the next. Potatoes are also an excellent storage crop.

Potato plants are herbaceous perennials that grow about 60 cm (24 in) high, depending on variety, with the culms dying back after flowering.

They bear white, pink, red, blue, or purple flowers with yellow stamens. In general, the tubers of varieties with white flowers have white skins, while those of varieties with coloured flowers tend to have pinkish skins. Potatoes are mostly cross-pollinated by insects such as bumblebees, which carry pollen from other potato plants, though a substantial amount of self-fertilizing occurs as well. Tubers form in response to decreasing day length, although this tendency has been minimized in commercial varieties.

Figure: *Potato plants*

After potato plants flower, some varieties produce small green fruits that resemble green cherry tomatoes, each containing up to 300 true seeds. Potato fruit contains large amounts of the toxic alkaloid solanine and is therefore unsuitable for consumption. All new potato varieties are grown from seeds, also called "true seed" or "botanical seed" to distinguish it from seed tubers. By finely chopping the fruit and soaking it in water, the seeds separate from the flesh by sinking to the bottom after about a day, while the remnants of the fruit float. Any potato variety can also be propagated vegetatively by planting tubers, pieces of tubers cut to include at least one or two eyes, or cuttings, a practice used in greenhouses for the production of healthy seed tubers. Some commercial potato varieties do not produce seeds at all as they bear imperfect flowers, and are instead propagated only from tuber pieces. These tubers or tuber pieces are called "seed potatoes," because the potato itself functions as "seed".

Genetics

There are about 5,000 potato varieties worldwide. Three thousand of them are found in the Andes alone, mainly in Peru, Bolivia, Ecuador, Chile, and Colombia. They belong to eight or nine species, depending on the taxonomic school. Apart from the 5,000 cultivated varieties, there are about 200 wild species and subspecies, many of which can be cross-bred with cultivated varieties. Cross-breeding has been done repeatedly to transfer resistances to certain pests and diseases from the gene pool of wild species to the gene pool of cultivated potato species. Genetically modified varieties have met public resistance in the United States and in the European Union.

The major species grown worldwide is *Solanum tuberosum* (a tetraploid with 48 chromosomes), and modern varieties of this species are the most widely cultivated. There are also four diploid species (with 24 chromosomes): *S. stenotomum*, *S. phureja*, *S. goniocalyx*, and *S. ajanhuiri*. There are two triploid species (with 36 chromosomes): *S. chaucha* and *S. juzepczukii*. There is one pentaploid cultivated species (with 60 chromosomes): *S. curtilobum*. There are two major subspecies of *Solanum tuberosum*: *andigena*, or Andean; and *tuberosum*, or Chilean. The Andean potato is adapted to the short-day conditions prevalent in the mountainous equatorial and tropical regions where it originated; the Chilean potato, however, native to the Chiloé Archipelago, is adapted to the long-day conditions prevalent in the higher latitude region of southern Chile.

The International Potato Centre, based in Lima, Peru, holds an ISO-accredited collection of potato germplasm. The international Potato Genome Sequencing Consortium announced in 2009 that they had achieved a draft sequence of the potato genome. The potato genome contains 12 chromosomes and 860 million base pairs, making it a medium-sized plant genome.

More than 99 percent of all current varieties of potatoes currently grown are direct descendants of a subspecies that once grew in the lowlands of south-central Chile. Nonetheless, genetic testing of the wide variety of cultivars and wild species affirms that all potato subspecies derive from a single origin in the area of present-day southern Peru and extreme northwestern Bolivia (from a species in the *Solanum brevicaule* complex).

Most modern potatoes grown in North America arrived through European settlement and not independently from the South American sources; however, at least one wild potato species, *Solanum fendleri*, is found as far north as Texas and is used in breeding for resistance to a nematode species that attacks cultivated potatoes. A secondary centre of genetic variability of the potato is Mexico, where important wild species that have been used extensively in modern breeding are found, such as the hexaploid *Solanum demissum*, as a source of resistance to the devastating late blight disease. Another relative native to this region, *Solanum bulbocastanum*, has been used to genetically engineer the potato to resist potato blight.

Potatoes yield abundantly with little effort, and adapt readily to diverse climates as long as the climate is cool and moist enough for the plants to gather sufficient water from the soil to form the starchy tubers. Potatoes do not keep very well in storage and are vulnerable to molds that feed on the stored tubers and quickly turn them rotten; however, crops such as grain can be stored for several years with a low risk of rot.

Growth and Cultivation

Potatoes are generally grown from seed potatoes – these are tubers specifically grown to be disease free and provide consistent and healthy plants. To be disease free, the areas where seed potatoes are grown are selected with care. In the USA this restricts production of seed potatoes to only 15 states out of the 50 states that grow potatoes. These locations are selected for their cold hard winters that kill pests and long sunshine hours in the summer for optimum growth. In the UK, most seed potatoes originate in Scotland in areas where westerly winds prevent aphid attack and thus prevent spread of potato virus pathogens. Potato growth has been divided into five phases.

During the first phase, sprouts emerge from the seed potatoes and root growth begins. During the second, photosynthesis begins as the plant develops leaves and branches. In the third phase stolons develop from lower leaf axils on the stem and grow downwards into the ground and on these stolons new tubers develop as swellings of the stolon. This phase is often (but not always) associated with flowering. Tuber formation halts when soil temperatures reach 27 °C (81 °F); hence potatoes are considered a cool-season crop. Tuber bulking occurs during the fourth phase, when the plant begins investing the majority of its resources in its newly formed tubers. At this stage, several factors are critical to yield: optimal soil moisture and temperature, soil nutrient availability and balance, and resistance to pest attacks. The final phase is maturation: The plant canopy dies back, the tuber skins harden, and their sugars convert to starches.

New tubers may arise at the soil surface. Since exposure to light leads to greening of the skins and the development of solanine, growers are interested in covering such tubers. Commercial growers usually address this problem by piling additional soil around the base of the plant as it grows ("hilling", or in British English "earthing up"). An alternative method used by home gardeners and smaller-scale growers involves covering the growing area with organic mulches such as straw or with plastic sheets.

Correct potato husbandry can be an arduous task in some circumstances. Good ground preparation, harrowing, plowing, and rolling are always needed, along with a little grace from the weather and a good source of water. Three successive plowings, with associated harrowing and rolling, are desirable before planting. Eliminating all root-weeds is desirable in potato cultivation. In general, the potatoes themselves are grown from the eyes of another potato and not from seed. Home gardeners often plant a piece of potato with two or three eyes in a hill of mounded soil. Commercial growers plant potatoes as a row crop using seed tubers, young plants or microtubers and may mound the entire row. Seed potato crops are 'rogued' in some countries to eliminate diseased plants or those of a different variety from the seed crop.

Potatoes are sensitive to heavy frosts, which damage them in the ground. Even cold weather makes potatoes more susceptible to bruising and possibly later rotting, which can quickly ruin a large stored crop.

At harvest time, gardeners usually dig up potatoes with a long-handled, three-prong "grape" (or graip), i.e., a spading fork, or a potato hook, which is similar to the graip but with tines at a 90° angle to the handle. In larger plots, the plow is the fastest implement for unearthing potatoes. Commercial harvesting is typically done with large potato harvesters, which scoop up the plant and surrounding earth. This is transported up an apron chain consisting of steel links several feet wide, which separates some of the dirt. The chain deposits into an area where further separation occurs. Different designs use different systems at this point. The most complex designs use vine choppers and shakers, along with a blower system or "Flying Willard" to separate the potatoes from the plant. The result is then usually run past workers who continue to sort out plant material, stones, and rotten potatoes before the potatoes are continuously delivered to a wagon or truck. Further inspection and separation occurs when the potatoes are unloaded from the field vehicles and put into storage.

Immature potatoes may be sold as "new potatoes" and are particularly valued for taste. These are often harvested by the home gardener or farmer by "grabbling", i.e. pulling out the young tubers by hand while leaving the plant in place.

Potatoes are usually cured after harvest to improve skin-set. Skin-set is the process by which the skin of the potato becomes resistant to skinning damage. Potato tubers may be susceptible to skinning at harvest and suffer skinning damage during harvest and handling operations. Curing allows the skin to fully set and any wounds to heal. Wound-healing prevents infection and water-loss from the tubers during storage. Curing is normally done at relatively warm temperatures 50 to 60 °C (122 to 140 °F) with high humidity and good gas-exchange if at all possible.

Potatoes always do best in full sun. They are aggressively rooting plants, and we find that they will produce the best crop when planted in a light, loose, well-drained soil. Potatoes prefer a slightly acid soil with a PH of 5.0 to 7.0. Fortunately potatoes are very adaptable and will almost always produce a respectable crop, even when the soil conditions and growing seasons are less than perfect. Always keep your potato patch weed-free for best results. Potatoes should be rotated in the garden, never being grown in the same spot until there has been a 3-4 year absence of potatoes. If you are new at growing potatoes, ask around; chances are there are many older gardeners in your area who have years of gardening experience.

Planting Times

Potatoes may be planted as soon as the ground can be worked in the early spring, but do keep soil temperatures in mind. Potato plants will not begin to grow until the soil temperature has reached 45 degrees F. The soil should be moist, but not water-logged. Potatoes can tolerate a light frost, but you should provide some frost protection for the plants if you know that a hard, late season freeze is coming. If you want to extend storage times, and have a long growing season, you can plant a second crop as late as June 15 and harvest the potatoes as late as possible.

Cutting Potatoes Before Planting

A week or two before your planting date, set your seed potatoes in an area where they will be exposed to light and temperatures between 60-70 degrees F – this will begin the sprouting process. A day or two before planting, use a sharp, clean knife to slice the larger seed potatoes into smaller pieces. Each piece must contain at least 1 or 2 eyes or buds and be no smaller than 1 inch in diametre. It is best to allow the cut surfaces to dry to prevent rotting once planted. Smaller potatoes may be planted whole – a good rule of thumb is to plant them whole if they are smaller in size than a golf ball.

Planting in the Garden

We find that potatoes are best grown in rows. To begin with, dig a trench that is 6-8 inches deep. Plant each piece of potato (cut side down, with the eyes pointing up) every 12-15 inches, with the rows spaced 3 feet apart. If your space is limited or if you would like to grow only baby potatoes, you can decrease the spacing between plants.

To begin with only fill the trench in with 4 inches of soil. Let the plants start to grow and then continue to fill in the trench and even mound the soil around the plants as they continue to grow. Prior to planting, always make sure to cultivate the soil one last time. This will remove any weeds and will loosen the soil and allow the plants to become established more quickly.

Water Supply

Keep your potato vines well watered throughout the summer, especially during the period when the plants are flowering and immediately following the flowering stage. During this flowering period the plants are creating their tubers and a steady water supply is crucial to good crop outcome. Potatoes do well with 1-2 inches of water or rain per week. When the foliage turns yellow and begins to die back, discontinue watering. This will help start curing the potatoes for harvest time.

Harvesting Your Potatoes

Baby potatoes typically can be harvested 2-3 weeks after the plants have finished flowering. Gently dig around the plants to remove potatoes for fresh eating, being careful not to be too intrusive. Try to remove the biggest new potatoes and leave the smaller ones in place so they can continue to grow. Only take what you need for immediate eating. Homegrown new potatoes are a luxury and should be used the same day that they are dug. Potatoes that are going to be kept for storage should not be dug until 2-3 weeks after the foliage dies back. Carefully dig potatoes with a sturdy fork and if the weather is dry, allow the potatoes to lay in the field, unwashed, for 2-3 days. This curing step allows the skins to mature and is essential for good storage. If the weather during harvest is wet and rainy, allow the potatoes to cure in a dry protected area like a garage or covered porch.

Storage Conditions

At Heritage Farm we are able to store potatoes well into the spring in our underground root cellar. Try to find a storage area that is well ventilated, dark, and cool. The ideal temperature is between 35 and 40 degrees F. Keep in mind that some varieties are better keepers than others. Varieties like Red Gold and Rose Gold are best used in the fall, and others like Carola and Russets are exceptional keepers.

Pests

The historically significant *Phytophthora infestans* (late blight) remains an ongoing problem in Europe and the United States. Other potato diseases include *Rhizoctonia, Sclerotinia,* black leg, powdery mildew, powdery scab and leafroll virus.

Insects that commonly transmit potato diseases or damage the plants include the Colorado potato beetle, the potato tuber moth, the green peach aphid (*Myzus persicae*), the potato aphid, beetleafhoppers, thrips, and mites. The potato root nematode is a microscopic worm that thrives on the roots, thus causing the potato plants to wilt. Since its eggs can survive in the soil for several years, crop rotation is recommended.

Jerusalem Artichokes

Potatoes aren't the only terrific tuber out there! Learn how to cultivate Jerusalem artichokes, which are knobby, nutty-flavoured tubers.

The Jerusalem artichoke (*Helianthus tuberosus*), also called sunroot, sunchoke, earth apple or *topinambour*, is a species of sunflower native to eastern North America, and found from eastern Canada and Maine west to North Dakota, and south to northern Florida and Texas. It is also cultivated widely across the temperate zone for its tuber, which is used as a root vegetable.

It is a herbaceous perennial plant growing to 1.5–3 m (4 ft 11 in–9 ft 10 in) tall with opposite leaves on the upper part of the stem but alternate below. The leaves have a rough, hairy texture and the larger leaves on the lower stem are broad ovoid-acute and can be up to 30 cm (12 in) long, and the higher leaves smaller and narrower.

The flowers are yellow and produced in capitate flowerheads, which are 5–10 cm (2.0–3.9 in) in diametre, with 10–20 ray florets.

The tubers are elongated and uneven, typically 7.5–10 cm (3.0–3.9 in) long and 3–5 cm (1.2–2.0 in) thick, and vaguely resembling ginger root in appearance, with a crisp texture when raw. They vary in colour from pale brown to white, red, or purple.

The artichoke contains about 10% protein, no oil, and a surprising lack of starch. However, it is rich in the carbohydrate inulin (76%), which is a polymer of the monosaccharide fructose.

Tubers stored for any length of time will convert their inulin into its component fructose. Jerusalem artichokes have an underlying sweet taste because of the fructose, which is about one and a half times sweeter than sucrose.

Jerusalem artichokes have also been promoted as a healthy choice for type 2 diabetics, because fructose is better tolerated by people who are type 2 diabetic. It has also been reported as a folk remedy for diabetes. Temperature variances have been shown to affect the amount of inulin the Jerusalem artichoke can produce. When not in tropical regions, it has been shown to make less inulin than when it is in a warmer region.

Etymology

Despite its name, the Jerusalem artichoke has no relation to Jerusalem, and it is not a type of artichoke, though both are members of the daisy family. The origin of the name is uncertain. Italian settlers in the USA called the plant *girasole*, the Italian word for sunflower, because of its resemblance to the garden sunflower (note: both the sunflower and the sunchoke are part of the same genus: *Helianthus*). Over time, the name *girasole* may have been changed to Jerusalem. The English later corrupted girasole artichoke (meaning, "sunflower artichoke") to Jerusalem artichoke. Another explanation for the name is that the Pilgrims, when they came to the New World, named the plant with regard to the "New Jerusalem" they believed they were creating in the wilderness. There have also been various other names applied to the plant, such as the French or Canada potato, topinambour, and lambchoke. Sunchoke, a name by which it is still known today, was invented in the 1960s by Frieda Caplan, a produce wholesaler who was trying to revive the plant's appeal.

The artichoke part of the Jerusalem artichoke's name comes from the taste of its edible tuber. Samuel de Champlain, the French explorer, sent the first samples of the plant to France, noting its taste was similar to an artichoke.

Handbuch des speziellen Gemüsebaus attributes the name "topinambur" to a 17th-century coincidence at the Vatican (in 1615) when a sample of the tuber from Canada was on display, as having helped French Canadian settlers to survive the Winter; at the same time a member of the Brazilian coastal tribe, the Tupinambá, was also visiting. The New World connection evidently resulted in the name for Jerusalem artichoke now used in French, German, Italian, Romanian, Russian, and Spanish.

History

Jerusalem artichokes were first cultivated by the Native Americans long before the arrival of the Europeans; this extensive cultivation obscures the exact native range of the species. The French explorer

Samuel de Champlain found domestically grown plants at Cape Cod in 1605. He then brought the plant back with him to France. By the mid-1600s, the Jerusalem artichoke had become a very common vegetable for human consumption in Europe and the Americas and was also used for livestock feed in Europe and colonial America. The French in particular were especially fond of the vegetable, which reached its peak popularity at the turn of the nineteenth century. The Jerusalem artichoke was titled 'best soup vegetable' in the 2002 Nice Festival For The Heritage Of The French Cuisine.

Cultivation and Uses

Figure: *Young plants in a garden*

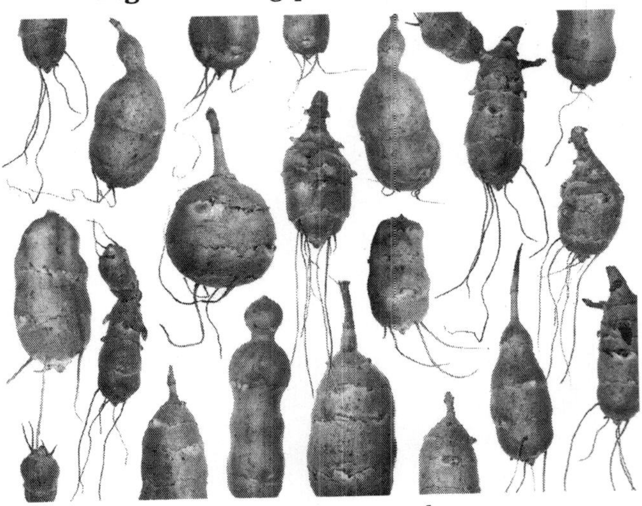

Figure: *Sunroot tubers*

Unlike most tubers, but in common with other members of the Asteraceae (including the artichoke), the tubers store the carbohydrate inulin (not to be confused with insulin) instead of starch. For this reason, Jerusalem artichoke tubers are an important source of inulin used as a dietary fibre in food manufacturing.

Crop yields are high, typically 16–20 tonnes/ha for tubers, and 18–28 tonnes/ha green weight for foliage. Jerusalem artichoke also has potential for production of ethanol fuel, using inulin-adapted strains of yeast for fermentation.

Jerusalem artichokes are easy to cultivate, which tempts gardeners to simply leave them completely alone to grow. However, the quality of the edible tubers degrades unless the plants are dug up and replanted in fertile soil. This can be a chore, as even a small piece of tuber will grow if left in the ground, making the hardy plant a potential weed. In fact, the plant can be pernicious. It can ruin gardens by smothering or overshadowing nearby plants and can overtake huge areas if left untamed. Commercial fields growing sunchoke which then change to other vegetables or crops often must be treated with glyphosate to stop the spread of the sunchokes. Each sunchoke root can make an additional 75 to as many as 200 tubers by fall end.

The tubers are sometimes used as a substitute for potatoes: they have a similar consistency, and in their raw form have a similar texture, but a sweeter, nuttier flavour; raw and sliced thinly, they are fit for a salad. The carbohydrates give the tubers a tendency to become soft and mushy if boiled, but they retain their texture better when steamed. The inulin cannot be broken down by the human digestive system, but is metabolized by bacteria in the colon. This can cause flatulence and, in some cases, gastric pain. *Gerard's Herbal*, printed in 1621, quotes the English planter John Goodyer on Jerusalem artichokes:

> *"which way soever they be dressed and eaten, they stir and cause a filthy loathsome stinking wind within the body, thereby causing the belly to be pained and tormented, and are a meat more fit for swine than men."*

Jerusalem artichokes have 650 mg potassium per 1 cup (150g) serving. They are also high in iron, and contain 10-12% of the US RDA of fibre, niacin, thiamine, phosphorus and copper.

Jerusalem artichokes can be used as animal feed, but they must be washed before being fed to most animals. However pigs can forage and safely eat them directly from the ground. The stalks and leaves can be harvested and used for silage, though cutting the tops greatly reduces the harvest of the roots.

Parsnips

Any gardener can be successful growing parsnips, a delicious root crop that tastes best when harvested in early winter, after the soil has turned cold.

The parsnip (*Pastinaca sativa*) is a root vegetable closely related to the carrot and parsley. It is a biennial plant usually grown as an annual. Its long tuberous root has cream-coloured skin and flesh and can be left in the ground when mature as it becomes sweeter in flavour after winter frosts. In its first growing season, the plant has a rosette of pinnate, mid-green leaves. If unharvested, it produces its flowering stem, topped by an umbel of small yellow flowers, in its second growing season. By this time the stem is woody and the tuber inedible. The seeds are pale brown, flat and winged.

The parsnip is native to Eurasia. It has been used as a vegetable since antiquity and was cultivated by the Romans, although there is some confusion in the literature of the time between parsnips and carrots. It was used as a sweetener before the arrival in Europe of cane sugar. It was introduced into the United States in the nineteenth century.

The parsnip is usually cooked but can also be eaten raw. It is high in vitamins and minerals, especially potassium. It also contains antioxidants and both soluble and insoluble dietary fibre. It can be cultivated in deep, stone-free soils and is attacked by the carrot fly and other insect pests, viruses and fungal diseases, of which canker is the most serious. In sunlight, handling the stems and foliage can cause a skin rash.

Description

Parsnips are grown for their fleshy, edible cream-coloured taproots. The roots are generally smooth, although lateral roots sometimes form. Most are cylindrical, but some cultivars have a more bulbous shape, which generally tend to be favoured by food processors as they are more resistant to breakage. The plant has a apical meristem that produces a rosette of pinnate leaves, each with several pairs of leaflets with toothed margins. The lower leaves have short stems, the upper ones are stemless, and the terminal leaves have three lobes. The highly branched floral stem is hollow and grooved, and can grow to more than 150 cm (60 in) tall.

Parsnip is a biennial with a rosette of roughly hairy leaves that has a pungent odour when crushed. The petioles are grooved and have sheathed bases. The leaves are once- or twice-pinnate with broad, ovate, sometimes lobed leaflets with toothed margins; they grow up to 40 cm (16 in) long. The flower stalk develops in the second year, growing to a height of 40 to 200 cm (20 to 80 in). It is hairy, grooved, hollow (except at the nodes), and sparsely branched. It has a few stalkless, single-lobed leaves measuring 5 to 10 cm (2 to 4 in) long that are arranged in opposite pairs.

The yellow flowers are in a loose, compound umbel measuring 10 to 20 cm (4 to 8 in) in diametre. There are 6–25 straight pedicels, each measuring 2–5 cm (1–2 in) that support the umbellets (secondary umbels). The umbels and umbellets usually have no upper or lower bracts. The flowers have tiny sepals or lack them entirely, and measure about 3.5 mm. They consist of five yellow petals that are curled inward, five stamens, and one pistil. The fruits, or schizocarps, are oval and flat, with narrow wings and short, spreading styles. They are coloured straw to light brown, and measure 4–8 mm long.

Despite the slight morphological differences between the two, wild parsnip is the same taxon as the cultivated version, and the two will readily cross-pollinate. Parsnip has a chromosome number of $2n=22$.

History

Like carrots, parsnips are native to Eurasia and have been eaten there since ancient times. Zohary and Hopf note that the archaeological evidence for the cultivation of the parsnip is "still rather limited", and that Greek and Roman literary sources are a major source about its early use. They warn that "there are some difficulties in distinguishing between parsnip and carrot (which, in Roman times, were white or purple) in classical writings since both vegetables seem to have been sometimes called *pastinaca* yet each vegetable appears to be well under cultivation in Roman times". The parsnip was much esteemed and the Emperor Tiberius accepted part of the tribute payable to Rome by Germany in the form of parsnips. In Europe, the vegetable was used as a source of sugar before sugar cane and beet were available. As *pastinache comuni*, the "common" *pastinaca* figures in the long list of comestibles enjoyed by the Milanese given by Bonvesin de la Riva in his "Marvels of Milan" (1288). This plant was introduced to North America simultaneously by the French colonists in Canada and the British in the Thirteen Colonies for use as a root vegetable, but in the mid-nineteenth century it was replaced as the main source of starch by the potato and consequently was less widely cultivated.

In 1859, a new cultivar called "Student" was developed by James Buckman at the Royal Agricultural College in England. He back-crossed cultivated plants to wild stock, aiming to demonstrate how native plants could be improved by selective breeding. This experiment was so successful that the Student became the major variety in cultivation in the late nineteenth century.

Taxonomy

- Like most plants of agricultural importance, several subspecies and varieties of *P. sativa* have been described, but these are mostly no longer recognized as independent taxa, but rather, morphological variations of the same taxon.
- var. *siamensis* Roem. & Schult. ex Alef.

In Eurasia, some authorities distinguish between cultivated and wild versions of parsnip by using subspecies *sylvestris* for the latter, or even elevating it to species status as *Pastinaca sylvestris*. In Europe, various subspecies have been named based on characteristics such as the hairiness of the leaves, the extent to which the stems are angled or rounded, and the size and shape of the terminal umbel. The etymology of the genus name *Pastinaca* is not known with certainty, but is probably derived from either the Latin word *pastino*, meaning "to prepare the ground for planting of the vine" or *pastus*, meaning "food". The specific epithet *sativa* means "sown".

Uses

Parsnips resemble carrots and can be used in similar ways but they have a sweeter taste, especially when cooked. While parsnips can be eaten raw, they are more commonly served cooked. They can be baked, boiled, pureed, roasted, fried or steamed. When used in stews, soups and casseroles they give a rich flavour.

In some cases, the parsnip is boiled and the solid portions are removed from the soup or stew, leaving behind a more subtle flavour than the whole root, and starch to thicken the dish. Roast parsnip is considered an essential part of Christmas dinner in some parts of the English-speaking world and frequently features in the traditional Sunday Roast. Parsnips can also be fried or thinly sliced and made into crisps. Parsnips can be made into a wine that has a taste similar to Madeira.

In Roman times, parsnips were believed to be an aphrodisiac. However, parsnips do not typically feature in modern Italian cooking. Instead, they are fed to pigs, particularly those bred to make Parma ham.

Nutritional Properties

A typical 100 g parsnip contains 75 Calories (230 kJ) of energy. Most parsnip cultivars consist of about 80% water, 5% sugar, 1% protein, 0.3% fat and 5% dietary fibre. The parsnip is rich in vitamins and minerals and is particularly rich in potassium with 375 mg per 100 g. Several of the B-group vitamins are present but most of the vitamin C is lost in cooking. Since most of the vitamins and minerals are found close to the skin many will be lost unless the root is finely peeled or cooked whole. During frosty weather, part of the starch is converted to sugar and the root tastes sweeter.

The consumption of parsnips has potential health benefits. They contain anti-oxidants such as falcarinol, falcarindiol, panaxydiol and methyl-falcarindiol which have anti-cancer, anti-inflammatory and anti-fungal properties. The dietary fibre in parsnips is partly of the soluble and partly the insoluble type and comprises cellulose, hemicellulose and lignin. The high fibre content of parsnips may help prevent constipation and reduce blood cholesterol levels.

Etymology

While folk etymology sometimes assumes the name is a portmanteau of parsley and turnip, it actually comes from Middle English *pasnepe*, alteration (influenced by *nep*, turnip) of Old French *pasnaie* (now *panais*) from Latin *pastinum,* a kind of fork, whose ending was changed to -*nip* by analogy with *turnip* because it was mistakenly assumed to be a kind of turnip.

Cultivation

The wild parsnip from which the modern cultivated varieties were derived is a plant of dry rough grassland and waste places, particularly on chalk and limestone. Parsnips are biennials but are normally grown as annuals. Sandy and loamy soils are preferable to silt, clay and stony ground as the latter produce short, forked roots.. Parsnip seed significantly deteriorates in viability if stored for long. Seeds are usually planted in early spring, as soon as the ground can be worked to a fine tilth, in the position where the plants are to grow. The growing plants are thinned and kept weed free. Harvesting begins in late fall after the first frost, and continues through winter. The rows can be covered with straw to enable the crop to be lifted during frosty weather. Low soil temperatures cause some of the starches stored in the roots to be converted into sugars, giving them a sweeter taste.

Cultivation Problems

Parsnip leaves are sometimes tunnelled by the larvae of the celery leaf miner (*Euleia heraclei*). Irregular pale brown passages can be seen meandering between the upper and lower surfaces of the leaves. The effects are most serious on young plants as whole leaves may shrivel and die. Treatment is by removing affected leaflets or whole leaves, or by chemical means.

The crop can be attacked by larvae of the carrot fly (*Chamaepsila rosae*). This pest feeds on the outer layers of the root, burrowing its way inside later in the season. Seedlings may be killed while larger roots are rendered useless. The damage done provides a point of entry for fungal rots and canker.

The fly is attracted by the smell of bruised tissue and damage can be minimized by sowing thinly, providing good growing conditions, avoiding thinning the plants and weeding carefully. Parsnip is used as a food plant by the larvae of some Lepidopteran species including the parsnip swallowtail, the common swift moth, the garden dart moth and the ghost moth. The larvae of the parsnip moth (*Depressaria radiella*), native to Europe and accidentally introduced to North America in the mid-1800s, construct their webs on the umbels, feeding on flowers and developing seeds.

Parsnip canker is a serious disease of this crop. Black or orange-brown patches occur around the crown and shoulders of the root accompanied by cracking and hardening of the flesh. It is more likely to occur when seed is sown into cold, wet soil, the pH of the soil is too low or the roots have already been damaged by carrot fly larvae. Better results may be obtained by choosing a resistant variety and sowing later in the season, producing smaller, more closely spaced plants. Several fungi are associated with canker, including *Phoma complanata*, *Ilyonectria radicicola*, *Itersonilia pastinaceae*, and *I. perplexans*. In Europe, *Centrospora acerina* has been found to cause a black rot that kills the plant early. Watery soft rot, caused by *Sclerotinia minor* and *S. sclerotiorum*, causes the taproot to become soft and watery. A white or buff-coloured mold grows on the surface. The pathogen is most common in temperate and subtropical regions that have a cool wet season.

Violet root rot caused by the fungus *Helicobasidium purpureum* sometimes affects the roots, covering them with a purplish mat to which soil particles adhere. The leaves become distorted and discoloured and the mycelium can spread through the soil between plants. Some weeds can harbour this fungus and it is more prevalent in wet, acid conditions. Affected plants should be removed and burnt and susceptible crops should not be grown in the area for four years. *Erysiphe heraclei* causes a powdery mildew that can cause significant crop loss. Infestation by this causes results in yellowing of the leaf and loss of foliage. Moderate temperatures and high humidity favour the development of the disease.

Several viruses are known to infect the plant, including seed-borne strawberry latent ringspot virus, parsnip yellow fleck virus, parsnip leafcurl virus, parsnip mosaic potyvirus, and potyvirus celery mosaic virus. The latter causes clearing or yellowing of those areas of the leaf immediately aligning the veins, the appearance of ochre mosaic spots, and leaf crinkling in infected plants.

Dangers

While the root of the parsnip is edible, handling the shoots and leaves of the plant requires caution as the sap is toxic. Like many other members of the family Apiaceae, the parsnip contains furanocoumarin, a photosensitive chemical that causes a condition known as phytophotodermatitis. The condition is a type of chemical burn rather than an allergic reaction, and is similar to the rash caused by poison ivy. Symptoms include redness, burning, and blisters. Afflicted areas can remain discoloured for up to two years. Although there have been some reports of gardeners experiencing toxic symptoms after coming into contact with foliage, these have been small in number compared to the number of people that grow the crop. The problem is most likely to occur on a sunny day when gathering foliage or pulling up old plants that have gone to seed. The symptoms have mostly been mild to moderate. The toxic properties of parsnip extracts are resistant to heating, or a storage period of several months. Toxic symptoms can also affect livestock and poultry in parts of their bodies where their skin is exposed.

Shallots

Fine-flavoured shallots can keep in storage for months, and growing shallots is easy. Whether you plant shallot cloves in fall or shallot seedlings in spring, by late summer you will be harvesting shallots.

The shallot (*Allium cepa* var. *aggregatum*, or the *A. cepa* Aggregatum Group) is a botanical variety of the species *Allium cepa*, to which the multiplier onion also belongs. The shallot was formerly classified as a separate species, *A. ascalonicum*, a name now considered a synonym of the currently accepted name. The genus *Allium*, which includes onions and garlic as well as shallots, is now classified in the plant family Amaryllidaceae, but was formerly considered to belong to the separate family Alliaceae.

Description and Cultivation

Like garlic, shallots are formed in clusters of offsets with a head composed of multiple cloves. The skin colour of shallots can vary from golden brown to gray to rose red, and their off-white flesh is usually tinged with green or magenta.

Shallots are extensively cultivated for culinary uses, propagated by offsets. In some regions ("long-season areas"), the offsets are usually planted in autumn (September or October in the Northern Hemisphere). In some other regions, the suggested planting time for the principal crop is early spring (typically in February or the beginning of March in the Northern Hemisphere).

In planting, the tops of the bulbs should be kept a little above ground, and the soil surrounding the bulbs is often drawn away when the roots have taken hold. They come to maturity in summer, although fresh shallots can now be found year-round in supermarkets. Shallots should not be planted on ground recently manured.

In Africa, shallots are grown in the area around Anloga in southeastern Ghana.

Culinary Uses

Shallots are used in fresh cooking in addition to being pickled. Finely sliced, deep-fried shallots are used as a condiment in Asian cuisine, often served with porridge. As a species of *Allium*, shallots taste somewhat like a common onion, but have a milder flavour. Like onions and garlic, when sliced, raw shallots release substances that irritate the human eye, resulting in production of tears.

Shallots appear to contain more flavonoids and phenols than other members of the onion genus.

Fresh shallots can be stored in cool, dry area (32 to 40 °F, 60 to 70% RH) for six months or longer. Chopped, dried shallots are also available.

Europe

Figure: *Shallots for sale in southern France*

In Europe, the Pikant, Atlas and Ed's Red types of shallots are the most common. In parts of southern France, the grey type is grown widely.

Asia

In Indian cuisines, the distinction between onions and shallots is weak; larger varieties of shallot are often confused with small red onions and used interchangeably. Indeed, most parts of India use the regional name for onion interchangeably with shallot (Maharashtra, for instance, where both are called *kanda*). The southern regions of India distinguish shallots from onions in recipes more often, especially the much loved tiny varieties (about the width of a finger); these are widely used in curries and different types of *sambar*, a lentil-based dish. Shallots pickled in red vinegar are common in many Indian restaurants, served along with sauces and papad on the condiments tray. Indians also use it as a home remedy for sore throats, mixed with jaggery or sugar. In Nepal, shallots are used as one of the ingredients for making *momo.* In Kashmir shallots are widely used in preparation of Wazwan Kashmiri cuisine, as they add distinct flavour and prevents curry from getting black which is a common problem with the onions.

In Iran, shallots, called *mousir*, are used in various ways, the most common being grated shallot mixed into dense yogurt, a combination served in almost every restaurant when one orders grills or kebabs. Shallots are also used to make different types of *torshi*, a sour Iranian side dish consisting of a variety of vegetables under vinegar, eaten with main dishes in small quantities. Shallot is also pickled -called *shour* in Persian- along with other vegetables to be served as *torshi.*

In Southeast Asian cuisines, such as those of Brunei, Vietnam, Thailand, Cambodia, Malaysia, Philippines, and Indonesia, both shallots and garlic (*bawang putih*, white onions) are often used as elementary spices. Raw shallot can also accompany cucumbers when pickled in mild vinegar solution. It is also often chopped finely, then fried until golden brown, resulting in tiny crispy shallot chips called *bawang goreng* (fried onions) in Indonesian language, which can be bought ready-made from groceries and supermarkets. Shallots enhance the flavour of many Southeast Asian dishes, such as fried rice variants. Crispy shallot chips are also used in southern Chinese cuisine. In Indonesia, shallots are sometimes made into pickles that are added to several traditional foods; the pickles' sourness is thought to increase one's appetite.

It is one of the pungent vegetables generally avoided by Buddhist vegetarians.

Taro

Taro is a common name for the corms and tubers of several plants in the Araceae family. Of these, *Colocasia esculenta* is the most widely cultivated and the subject of this article. More specifically, this article describes the "dasheen" form of taro; another variety of taro is known as *eddoe.*

Taro is native to Southern India and Southeast Asia. It is a perennial, tropical plant primarily grown as a root vegetable for its edible starchy corm, and as a leaf vegetable. It is a food staple in African, Oceanic and South Indian cultures and is believed to have been one of the earliest cultivated plants. *Colocasia* is thought to have originated in the Indo-Malayan region, perhaps in eastern India and Bangladesh, and spread eastward into Southeast Asia, eastern Asia, and the Pacific islands; westward to Egypt and the eastern Mediterranean; and then southward and westward from there into East Africa and West Africa, whence it spread to the Caribbean and Americas. It is known by many local names and often referred to as "elephant ears" when grown as an ornamental plant.

Cultivation

Taro can be grown in paddy fields where water is abundant or in upland situations where water is supplied by rainfall or supplemental irrigation. Taro is one of the few crops (along with rice and lotus) that can be grown under flooded conditions. This is due to air spaces in the petiole, which permit underwater gaseous exchange with the atmosphere. For a maximum dissolved oxygen supply, the water should be cool and flowing. Warm, stagnant water causes basal rotting. For maximum yields, the water level should be controlled so that the base of the plant is always under water.

Flooded cultivation has some advantages over dry-land cultivation: higher yields (about double), out-of-season production (which may result in higher prices), and weed control (which flooding facilitates). On the other hand, in flooded production systems taro requires a longer maturation period, investment in infrastructure, and higher operational costs, and monoculture is likely.

Like most root crops, taro and eddoes do well in deep, moist or even swampy soils where the annual rainfall exceeds 2,500 mm. Eddoes are more resistant to drought and cold. The crop attains maturity within six to twelve months after planting in dry-land cultivation and after twelve to fifteen months in wetland cultivation. The crop is harvested when the plant height decreases and the leaves turn yellow. These signals are usually less distinct in flooded taro cultivation.

Harvesting is usually done by hand tools, even in mechanized production systems. First, the soil around the corm is loosened, and then, the corm is pulled up by grabbing the base of the petioles. The global average yield is 6.2 tones/hectare but varies according to the region. In Asia, average yields reach 12.6 tones/hectare.

Toxicity

The plant is inedible when raw and considered toxic due to the presence of calcium oxalate crystals, typically as raphides. The toxin is minimized by cooking, especially with a pinch of baking soda. It can also be reduced by steeping taro roots in cold water overnight. Calcium oxalate is highly insoluble and contributes to kidney stones. It has been recommended to consume milk or other calcium-rich foods together with taro.

Culinary Use

The corms, which have a light purple colour due to phenolic pigments, are roasted, baked or boiled, and the natural sugars give a sweet nutty flavour. The starch is easily digestible, and since the grains are fine and small it is often used for baby food. Young taro leaves and stems can be eaten after boiling twice to remove the acrid flavour and the leaves are a good source of vitamins A and C and contain more protein than the corms.

Horseradish

Growing horseradish is possible in a wide range of climates because they are such tough, persistent plants. Horseradish roots are harvested from fall through winter, providing plenty of warmth to winter meals.

Horseradish (*Armoracia rusticana*, syn. *Cochlearia armoracia*) is a perennial plant of the Brassicaceae family (which also includes mustard, wasabi, broccoli, and cabbage). The plant is probably native to southeastern Europe and western Asia. It is now popular around the world. It grows up to 1.5 metres (4.9 feet) tall, and is cultivated primarily for its large, white, tapered root.

The intact horseradish root has hardly any aroma. When cut or grated, however, enzymes from the now-broken plant cells break down sinigrin (a glucosinolate) to produce allyl isothiocyanate (mustard oil), which irritates the mucous membranes of the sinuses and eyes. Grated mash should be used immediately or preserved in vinegar for best flavour. Once exposed to air or heat it will begin to lose its pungency, darken in colour, and become unpleasantly bitter tasting over time.

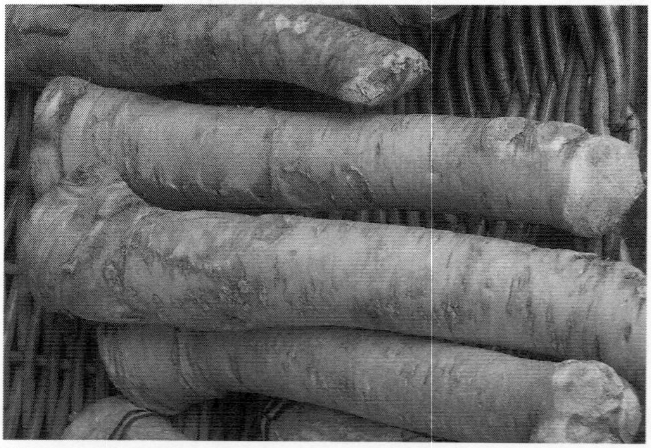

Figure: *Sections of roots of the horseradish plant*

Ultivation

Horseradish is perennial in hardiness zones 2–9 and can be grown as an annual in other zones, although not as successfully as in zones with both a long growing season and winter temperatures cold enough to ensure plant dormancy. After the first frost in the autumn kills the leaves, the root is dug and divided. The main root is harvested and one or more large offshoots of the main root are replanted to produce next year's crop. Horseradish left undisturbed in the garden spreads via underground shoots and can become invasive. Older roots left in the ground become woody, after which they are no longer culinarily useful, although older plants can be dug and re-divided to start new plants. The early season leaves can be distinctively different, asymmetric spiky, before the mature typical flat broad leaves start to be developed.

Pests and Diseases

Widely introduced by accident, "cabbageworms", the larvae of *Pieris rapae*, the Small White Butterfly, are a common caterpillar pest in horseradish. The adults are white butterflies with black spots on the forewings that are commonly seen flying around plants during the day.

The caterpillars are velvety green with faint yellow stripes running lengthwise down the back and sides. Full grown caterpillars are about 1-inch (25 mm) in length. They move sluggishly when prodded. They overwinter in green pupal cases. Adults start appearing in gardens after the last frost and are a problem through the remainder of the growing season. There are three to five overlapping generations a year. Mature caterpillars chew large, ragged holes in the leaves leaving the large veins intact. Handpicking is an effective control strategy in home gardens.

Culinary uses

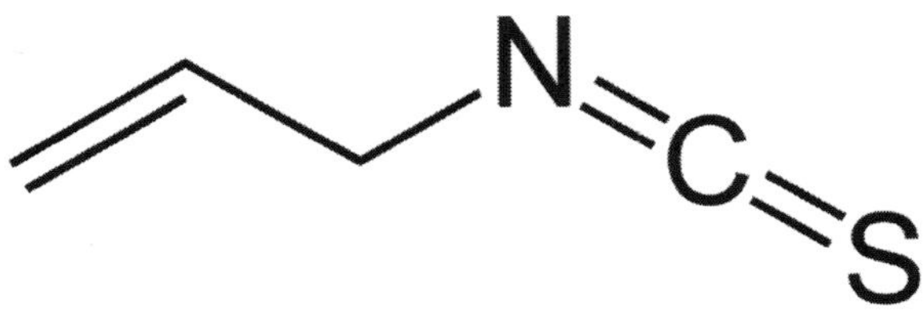

***Figure:** Allyl isothiocyanate is the pungent ingredient in fresh horseradish sauce.*

Cooks use the terms "horseradish" or "prepared horseradish" to refer to the grated root of the horseradish plant mixed with vinegar. Prepared horseradish is white to creamy-beige in colour. It will keep for months refrigerated but eventually will darken, indicating it is losing flavour and should be replaced. The leaves of the plant, while edible, are not commonly eaten, and are referred to as "horseradish greens", which have a flavour of root.

Vegetable

In Central and Eastern Europe horseradish is called *khren* (in various spellings like *kren*) in many Slavic languages, in Austria, in parts of Germany (where the other German name *Meerrettich* isn't used), in North-East Italy, and in Yiddish (*ëøééï* translitered as *khren*).

There are two varieties of khreyn. "Red" khreyn is mixed with red beet (beetroot) and "white" khreyn contains no beet. It is popular in Ukraine (under the name of *õð³í, khrin*), in Poland (under the name of *chrzan*), in Lithuania (krienai) in the Czech Republic (*køen*), in Russia (*õðåí, khren*), in Hungary (*torma*), in Romania (*hrean*), in Bulgaria (*õðÿí, khryan*), and in Slovakia (under the name of *chren*). Having this on the table is a part of Christian Easter and Jewish Passover tradition in Eastern and Central Europe.

- In parts of Southern Germany like Franconia, "Kren" is an essential component of the traditional wedding dinner. It is served with cooked beef and a dip made from lingonberry to balance the slight hotness of the Kren.

- In Poland, a variety with red beet is called *æwik³a z chrzanem* or simply æwik³a.
- In Ashkenazi European Jewish cooking beet horseradish is commonly served with gefilte fish.
- In Transylvania, Red beet with horseradish is also used as a salad served with lamb dishes at Easter called *sfecla cu hrean* and other Romanian regions.
- In Serbia, *ren* is an essential condiment with cooked meat and freshly roasted suckling pig.
- In Croatia, freshly grated horseradish (Croatian: Hren) is often eaten with boiled ham or beef.
- In Slovenia, and in the adjacent Italian regions of Friuli Venezia Giulia and nearby Italian region of Veneto, Horseradish (often grated and mixed with sour cream, vinegar, hard-boiled eggs, or apples) is also a traditional Easter dish.
- Further west in the Italian regions of Lombardy, Emilia-Romagna, and Piedmont, it is called "barbaforte (strong beard)" and is a traditional accompaniment to bollito misto; while in north-eastern regions like Trentino-Alto Adige/Südtirol, Veneto and Friuli-Venezia Giulia, it is still called "kren" or "cren".
- Horseradish is also used as a main ingredient for soups. In the Polish region of Silesia, horseradish soup is a common Easter Day dish.

Relation to Wasabi

The Japanese condiment wasabi, although traditionally prepared from the wasabi plant, is now usually made with horseradish due to the scarcity of the wasabi plant. The Japanese botanical name for horseradish is *seiyôwasabi*, or "Western wasabi". Both plants are members of the family Brassicaceae.

Nutritional and Biomedical Uses

Compounds found in horseradish have been widely studied for a plethora of health benefits. Horseradish contains volatile oils, notably mustard oil, which has antibacterial properties due to the presence of allyl isothiocyanate. Fresh, the plant also contains average 79.31 mg of vitamin C per 100 g of raw horseradish.

The enzyme horseradish peroxidase (HRP), found in the plant, is used extensively in molecular biology and biochemistry.

Nelumbo Nucifera

Nelumbo nucifera, known by numerous common names including Indian lotus, sacred lotus, bean of India, or simply lotus, is one of two species of aquatic plant in the family Nelumbonaceae. The Linnaean binomial *Nelumbo nucifera* (Gaertn.) is the currently recognized name for this species, which has been classified under the former names, *Nelumbium speciosum* (Willd.) and *Nymphaea nelumbo*, among others. Names other than *Nelumbo nucifera* (Gaertn.) are obsolete synonyms and should not be used in current works. This plant is an aquatic perennial. Under favourable circumstances its seeds may remain viable for many years, with the oldest recorded lotus germination being from that of seeds 1,300 years old recovered from a dry lakebed in northeastern China.

A common misconception is referring to the lotus as a water lily (*Nymphaea*), an entirely different plant, as can be seen in the centre of the flowers. *Nymphaea* lacks the structure that goes on to form the distinctive circular seed pod in the *Nelumbo nucifera*. Native to Tropical Asia and Queensland, Australia, it is commonly cultivated in water gardens. It is also the national flower of India and Vietnam.

Botany

The roots of *lotus* are planted in the soil of the pond or river bottom, while the leaves float on top of the water surface or are held well above it. The flowers are usually found on thick stems rising several centimetres above the leaves. The plant normally grows up to a height of about 150 cm and a horizontal spread of up to 3 metres, but some unverified reports place the height as high as over 5 metres. The leaves may be as large as 60 cm in diametre, while the showy flowers can be up to 20 cm in diametre.

Researchers report that the lotus has the remarkable ability to regulate the temperature of its flowers to within a narrow range just as humans and other warmblooded animals do. Dr. Roger S. Seymour and Dr. Paul Schultze-Motel, physiologists at the University of Adelaide in Australia, found that lotus flowers blooming in the Adelaide Botanic Gardens maintained a temperature of 30–35 °C (86–95 °F), even when the air temperature dropped to 10 °C (50 °F). They suspect the flowers may be doing this to attract coldblooded insect pollinators. The study, published in the journal *Nature*, is the latest discovery in the field of thermoregulation, heat-producing, plants. Two other species known to be able to regulate their temperature include *Symplocarpus foetidus* and *Philodendron selloum*.

An individual lotus can live for over a thousand years and has the rare ability to revive into activity after stasis. In 1994, a seed from a sacred lotus, dated at roughly 1,300 years old ± 270 years, was successfully germinated.

As mentioned earlier, the traditional Sacred Lotus is only distantly related to *Nymphaea caerulea*, but possesses similar chemistry. Both *Nymphaea caerulea* and *Nelumbo nucifera* contain the alkaloids nuciferine and aporphine.

The genome of the sacred lotus was sequenced in May 2013.

Uses

Figure: *Boiled, sliced lotus roots used in various Asian cuisines*

The distinctive dried seed heads, which resemble the spouts of watering cans, are widely sold throughout the world for decorative purposes and for dried flower arranging.

The flowers, seeds, young leaves, and "roots" (rhizomes) are all edible. In Asia, the petals are sometimes used for garnish, while the large leaves are used as a wrap for food, not frequently eaten (for example, as a wrapper for *zongzi*). In Korea, the leaves and petals are used as a tisane. Lotus rootlets are often pickled with rice vinegar, sugar, chili and/or garlic. It has a crunchy texture with sweet-tangy flavours. In Asian cuisine, it is popular with salad, prawns, sesame oil and/or coriander leaves. Lotus roots have been found to be rich in dietary fibre, vitamin C, potassium, thiamin, riboflavin, vitamin B_6, phosphorus, copper, and manganese, while very low in saturated fat.

The stamens can be dried and made into a fragrant herbal tea called *liánhuâ cha* (î,,±,6*f*) in Chinese, or (particularly in Vietnam) used to impart a scent to tea leaves. This Vietnamese lotus tea is called *trà sen*, *chè sen*, or *chè ýÛp sen*. The lotus seeds or nuts in Chinese) are quite versatile, and can be eaten raw or dried and popped like popcorn, *phool makhana*. They can also be boiled until soft and made into a paste, or boiled with dried longans and rock sugar to make a *tong sui* (sweet soup).

Combined with sugar, lotus seed paste becomes one of the most common ingredients used in pastries such as mooncakes, *daifuku*, and rice flour pudding.

In South Indian states, the Lotus Stem is sliced, marinated with salt to dry, and the dried slices are fried and used as a side dish. In Kerala and Tamil Nadu, this end product is called " Thamara Vathal". In Sri Lanka, the sliced Lotus Stem curry is a popular dish called "Nelum Ala".

A unique fabric from the lotus plant fibres is produced only at Inle lake, Union of Myanmar and is used for weaving special robes for Buddha images called *kya thingahn* (lotus robe).

Kohlrabi

Kohlrabi (German turnip or turnip cabbage) (*Brassica oleracea* Gongylodes group) is an annual vegetable, and is a low, stout cultivar of cabbage. Kohlrabi can be eaten raw as well as cooked.

Kohlrabi has been created by artificial selection for lateral meristem growth (a swollen, nearly spherical shape); its origin in nature is the same as that of cabbage, broccoli, cauliflower, kale, collard greens, and Brussels sprouts: they are all bred from, and are the same species as the wild cabbage plant (*Brassica oleracea*).

The taste and texture of kohlrabi are similar to those of a broccoli stem or cabbage heart, but milder and sweeter, with a higher ratio of flesh to skin. The young stem in particular can be as crisp and juicy as an apple, although much less sweet.

Except for the *Gigante* cultivar, spring-grown kohlrabi much over 5 cm in size tend to be woody, as do full-grown kohlrabi much over perhaps 10 cm in size; the Gigante cultivar can achieve great size while remaining of good eating quality. The plant matures in 55–60 days after sowing. Approximate weight is 150 g and has good standing ability for up to 30 days after maturity.

There are several varieties commonly available, including White Vienna, Purple Vienna, Grand Duke, Gigante (also known as "Superschmelz"), Purple Danube, and White Danube. Coloration of the purple types is superficial: the edible parts are all pale yellow. The leafy greens can also be eaten.

Preparation and Use

Kohlrabi stems are surrounded by two distinct fibrous layers that do not soften appreciably when cooked. These layers are generally peeled away prior to cooking or serving raw, with the result that the stems often provide a smaller amount of food than one might assume from their intact appearance.

The Kohlrabi root is frequently used raw in salad or slaws. It has a texture similar to that of a broccoli stem, but with a flavour that is sweeter and less vegetal.

Kohlrabi leaves are edible and can be used interchangeably with collard and kale.

Kohlrabi is an important part of the Kashmiri diet and one of the most commonly cooked foods. It is prepared with its leaves and served with a light gravy and eaten with rice.

Some varieties are grown as feed for cattle.

Pest Prevention in Root Vegetables

Preventive pest control is important when growing root veggies; when there are signs of pests, the damage may already have been done below ground. A yearly application of beneficial nematodes will help with wireworms, cutworms, and other soil-dwelling grubs. Apply a row cover to prevent flying pests, such as carrot rust fly, from invading and laying eggs. Be quick about replacing the cover after thinning young plants, for the smell of the bruised foliage is what attracts the pests. Laying down 1x2s along the edge of the floating row cover will help. If you staple the edges onto the laths, you can easily pick up the side to work on the patch and replace it as soon as you are finished. To keep out the female carrot fly, try wrapping a curtain of any material (fleece; plastic; tight mesh; row cover) onto stakes around the carrot patch. Ensure the material is tight to the ground.

Even if you don't see any pests, it's a good idea to have a few cards around to monitor the situation. Periodically check the cards, identify any pests and then find an organic solution. Wash all root crops under warm water using a plastic pot scrubber dedicated to vegetables. For extra cleansing dip washed vegetables into a lemon-water bath.

7

Some Other Important Vegetables

Rhubarb

Rhubarb (*Rheum rhabarbarum*) is a species of plant in the family Polygonaceae. They are herbaceous perennials growing from short, thick rhizomes. They have large leaves that are somewhat triangular, with long fleshy petioles. They have small flowers grouped in large compound leafy greenish-white to rose-red inflorescences.

In culinary use, fresh raw petioles (leaf stalks) are crisp (similar to celery) with a strong, tart taste. Most commonly, the plant's leaf stalks are cooked with sugar and used in pies and other desserts. A number of varieties have been domesticated for human consumption, most of which are recognised as *Rheum* x *hybridum* by the Royal Horticultural Society.

Rhubarb is usually considered a vegetable. In the United States, however, a New York court decided in 1947 that since it was used in the United States as a fruit, it counted as a fruit for the purposes of regulations and duties. A side effect was a reduction on imported rhubarb tariffs, as tariffs were higher for vegetables than fruits.

Rhubarb also contains glycosides—especially rhein, glucorhein, and emodin, which impart cathartic and laxative properties. It is hence useful as a cathartic in case of constipation.

Cultivation

Rhubarb is grown in many areas. With greenhouse production, it is available throughout much of the year. Rhubarb grown in hothouses (heated greenhouses) is called hothouse rhubarb, and is typically made available at consumer markets in early spring, before outdoor cultivated rhubarb is available. Hothouse rhubarb is usually brighter red, more tender and sweeter-tasting than outdoor rhubarb. In temperate climates, rhubarb is one of the first food plants harvested, usually in mid- to late spring (April/May in the Northern Hemisphere, October/November in the Southern Hemisphere, and the season for field-grown plants lasts until September. In the northwestern US states of Oregon and Washington, typically two harvests are from late April to May and from late June into July. Rhubarb is ready to consume as soon as harvested, and freshly cut stalks are firm and glossy.

In the United Kingdom, the first rhubarb of the year is harvested by candlelight in forcing sheds - where all other light is excluded - a practice that produces a sweeter, more tender stalk. These sheds are dotted around the noted "Rhubarb Triangle" of Wakefield, Leeds, and Morley.

Rhubarb grows year-round in warm climates, but in temperate climates, the above ground portion of the plant completely withers away at the onset of freezing temperature. The plant grows from the root at the return of warm weather. Rhubarb growth can be 'forced' or encouraged to grow early by raising the local temperature, usually by placing an upturned bucket over the new shoots.

Because rhubarb is a seasonal plant, obtaining fresh rhubarb out of season is difficult in colder climates, such as in the UK, Ireland, Russia, etc. Rhubarb thrives in areas of direct sunlight. Rhubarb can successfully be planted in containers, if they are large enough to accommodate a season's growth.

Rhubarb damaged by severe cold should not be eaten, as it may be high in oxalic acid, which migrates from the leaves and can cause illness.

The colour of rhubarb stalks can vary from the commonly associated crimson red, through speckled light pink, to simply light green. Rhubarb stalks are poetically described as "crimson stalks". The colour results from the presence of anthocyanins, and varies according to both rhubarb variety and production technique. The colour is not related to its suitability for cooking: The green-stalked rhubarb is more robust and has a higher yield, but the red-coloured stalks are much more popular with consumers.

Historical Cultivation

Rhubarb has been used for medical purposes by the Chinese for thousands of years, and appears in *The Divine Farmer's Herb-Root Classic*, which legend attributes to the mythical Shen Nung, the Yan Emperor, but is thought to have been compiled about 2700 years ago. Though Dioscurides' description of ñçïí or ñÜ, a root brought to Greece from beyond the Bosphorus may have included rhubarb, the commerce did not become securely established until Islamic times, when it was imported along the Silk Road, reaching Europe in the 14th century through the ports of Aleppo and Smyrna, and becoming known as "Turkish rhubarb". Later, when the usual route lay through Russia, "Russian rhubarb" became the familiar term.

For centuries, the plant has grown wild along the banks of the River Volga, for which the ancient Scythian hydronym was *Rhâ*. The expense of transportation across Asia made rhubarb expensive in medieval Europe—it was several times the price of other valuable herbs and spices, such as cinnamon, opium, and saffron. The merchant explorer Marco Polo therefore wanted to find where the plant was grown and harvested in the mountains of Tangut province. The value of rhubarb can be seen in Ruy Gonzáles de Clavijo's report of his embassy in 1403-05 to Timur in Samarkand: "The best of all merchandise coming to Samarkand was from China: especially silks, satins, musk, rubies, diamonds, pearls, and rhubarb..."

The term "rhubarb" is a combination of the Ancient Greek *rha* and *barbarum*; *rha* refers both to the plant and to the River Volga. Rhubarb first came to the United States in the 1820s, entering the country in Maine and Massachusetts and moving westwards with the European American settlers.

Uses

Rhubarb is grown primarily for its fleshy stalks, technically known as petioles. The use of rhubarb stems as food is a relatively recent innovation, first recorded in 17th century England, after affordable sugar became available to common people, and reaching a peak between the 20th century's two world wars.

Commonly, it is stewed with sugar or used in pies and desserts, but it can also be put into savory dishes or pickled. Rhubarb can be dehydrated and infused with fruit juice. In most cases, it is infused with strawberry juice to mimic the popular strawberry rhubarb pie.

Rhubarb root produces a rich brown dye similar to walnut husks. It is used in northern regions where walnut trees do not survive.

Cooking

For cooking, the stalks are often cut into 1 inch (2.5 cm) pieces and stewed (boiled in water). It is necessary only to barely cover the stalks with water, as rhubarb stalks contain a great deal of water on their own; $^{1}D_{2}$ to $^{3}D_{4}$ cup of sugar is added for each pound of rhubarb. Spices such as cinnamon and/or nutmeg can be added to taste. Sometimes a tablespoon of lime juice or lemon juice is added. The sliced stalks are boiled until soft. An alternative method is to simmer slowly without adding water, letting the rhubarb cook in its own juice.

At this stage, either alone or cooked with strawberries or apples as a sweetener, or with stem or root ginger, rhubarb can be used to make jam. Other fruits, with the addition of pectin (or using sugar with pectin already added), can be added to rhubarb to make a variety of jams.

A sauce (to which dried fruit can be added near the end) can be made by cooking further, until the sauce is mostly smooth and the remaining discrete stalks can easily be pierced with a fork, yielding a smooth tart-sweet sauce with a flavour similar to sweet and sour sauce. This sauce, rhubarb sauce, is analogous to applesauce. Like applesauce, this sauce is usually stored in the refrigerator and eaten cold. The sauce, after cornstarch mixed with water has been added to thicken it, may be used as filling for rhubarb pie, tarts, and crumbles. Sometimes stewed strawberries are mixed with the rhubarb to make strawberry-rhubarb pie. This common use has led to the slang term for rhubarb, *pie plant*, by which name it was more commonly known in the United States in the late 19th century. In her novella *The First Four Years*, American author Laura Ingalls Wilder refers to rhubarb as "pie plant". It can also be used to make a fruit wine.

In former days, a common and affordable sweet for children in parts of the United Kingdom and Sweden was a tender stick of rhubarb, dipped in sugar. It is still eaten this way in western Finland and Norway, Iceland and some other parts of the world, including rural eastern Ontario. In Chile, Chilean rhubarb is sold on the street with salt or dried chili pepper, not sugar.

Medicine

In traditional Chinese medicine, rhubarb roots have been used as a laxative for several thousand years. Rhubarb also appears in medieval Arabic and European prescriptions.

Chemistry

The roots and stems are rich in anthraquinones, such as emodin and rhein. These substances are cathartic and laxative, explaining the sporadic use of rhubarb as a dieting aid. The anthraquinone compounds have been separated from powdered rhubarb root for medicinal purposes.

The rhizomes ('roots') contain stilbenoid compounds (including rhaponticin), which seem to lower blood glucose levels in diabetic mice.

It also contains the flavanol glucosides (+)-catechin-5-O-glucoside and (-)-catechin-7-O-glucoside.

Mushroom

Edible mushrooms are the fleshy and edible fruit bodies of several species of macrofungi (fungi which bear fruiting structures that are large enough to be seen with the naked eye). They can appear either below ground (hypogeous) or above ground (epigeous) where they may be picked by hand. Edibility may be defined by criteria that include absence of poisonous effects on humans and desirable taste and aroma.

Edible mushrooms are consumed by humans as comestibles for their nutritional value and they are occasionally consumed for their supposed medicinal value. Mushrooms consumed by those practicing folk medicine are known as medicinal mushrooms. While hallucinogenic mushrooms (e.g. Psilocybin mushrooms) are occasionally consumed for recreational or religious purposes, they can produce severe nausea and disorientation, and are therefore not commonly considered edible mushrooms.

Edible mushrooms include many fungal species that are either harvested wild or cultivated. Easily cultivatable and common wild mushrooms are often available in markets, and those that are more difficult to obtain (such as the prized truffle and matsutake) may be collected on a smaller scale by private gatherers. Some preparations may render certain poisonous mushrooms fit for consumption.

Before assuming that any wild mushroom is edible, it should be identified. Proper identification of a species is the only safe way to ensure edibility. Some mushrooms that are edible for most people can cause allergic reactions in some individuals, and old or improperly stored specimens can cause food poisoning.

Deadly poisonous mushrooms that are frequently confused with edible mushrooms and responsible for many fatal poisonings include several species of the *Amanita* genus, in particular, *Amanita phalloides*, the *death cap*. Mushrooms growing in polluted locations can accumulate pollutants such as heavy metals.

A fraction of the many fungi consumed by humans are currently cultivated and sold commercially. Commercial cultivation is important ecologically, as there have been concerns of depletion of larger fungi such as chanterelles in Europe, possibly because the group has grown so popular yet remains a challenge to cultivate.

Commercially Cultivated

Figure: *Commercial cultivated Japanese edible mushroom species. Clockwise from left, enokitake, buna-shimeji, bunapi-shimeji, king oyster mushroom and shiitake.*

Mushroom cultivation has a long history, with over twenty species commercially cultivated. Mushrooms are cultivated in at least 60 countries with China, the United States, Netherlands, France and Poland being the top five producers in 2000.

- Some species are difficult to cultivate; others (particularly mycorrhizal species) have not yet been successfully cultivated. Some of these species are harvested from the wild, and can be found in markets. When in season they can be purchased fresh, and many species are sold dried as well.

Other Edible Wild Species

Many wild species are consumed around the world. The species which can be identified “in the field” (without use of special chemistry or a microscope) and therefore safely eaten vary widely from country to country, even from region to region. This list is a sampling of lesser-known species that are reportedly edible.

Fungiculture

Fungiculture is the process of producing food, medicine, and other products by the cultivation of mushrooms and other fungi.

Techniques

All mushroom growing techniques require the correct combination of humidity, temperature, substrate (growth medium) and inoculum (spawn or starter culture). Wild harvests, outdoor log inoculation and indoor trays all provide these elements.

Wild Harvesting

Due to its climate, the Pacific Northwest of the USA produces commercially valuable mushrooms. Mushroom gatherers have few requirements to begin business. Gatherers only need to supply funds for possible park fees, knowledge for identifying mushrooms and gathering time. There are significant disadvantages to relying on natural mushroom production. These sales may be unregulated, placing buyers at risk for buying toxic or inedible mushrooms. By honest error, harvests may include toxic or inedible species. No controls exist to regulate the quality or frequency of harvests, since gatherers rely on favourable natural conditions and weather to produce fruiting. Conflicts may arise between competing gatherers trying to harvest from the same location.

State parks in the Pacific Northwest or elsewhere may charge fees for mushroom gathering permits. Appalachia also produces edible wild mushrooms, including chanterelles and morels. Pickers may sell directly to distributors, restaurants, or sell their harvest through roadside stands wherever a natural supply of mushrooms is plentiful.

While there may be concern that harvesting wild mushrooms may exploit or damage a natural environment, harvesting wild mushrooms is different from harvesting wild plants, fishing or hunting animals. In these last three cases, removing individuals decreases the ability of a wild population to reproduce, since fewer adults remain. Removing adults leaves fewer individuals capable of reproducing and reduces genetic diversity.

Harvesting wild mushrooms removes only fruiting bodies and their attached spores. However, the fruiting bodies (mushrooms) have likely dropped spores before harvest time, or will likely drop them en route to the harvester's destination, further expanding the fungi's habitat. Arguably, the practice of mushroom harvesting may actually help the species being harvested. While truffles also represent the fruiting body of a larger underground network, they are an exception, since they rely on animal spore dispersion.

Additionally, reproduction and propagation can still occur by propagation of the parent mycelium. Harvesting removes none of the parent mycelium, which remains intact underground.

Outdoor Logs

Mushrooms can be grown on logs placed outdoors in stacks or piles, as has been done for hundreds of years. Sterilization is not performed in this method. Since production may be unpredictable and seasonal, less than 5% of commercially sold mushrooms are produced this way. Here, tree logs are inoculated with spawn, then allowed to grow as they would in wild conditions. Fruiting, or pinning, is triggered by seasonal changes, or by briefly soaking the logs in cool water. Shiitake and oyster mushrooms have traditionally been produced using the outdoor log technique, although controlled techniques such as indoor tray growing or artificial logs made of compressed substrate have been substituted.

Shiitake mushrooms grown under a forested canopy are considered non-timber forest products In the Northeast shiitake mushrooms can be cultivated on a variety of hardwood logs including oak, American beech, sugar maple and hophornbeam. Softwood should not be used to cultivate shiitake mushrooms. The resin of softwoods will often times inhibit the growth of the shiitake mushroom making it impractical as a growing substrate.

In order to produce shiitake mushrooms, 3 foot hardwood logs with a diametre ranging between 4 and 6 inches are inoculated with the mycelium of the shiitake fungus. Inoculation is completed by drilling holes in hardwood logs, filling the holes with cultured shiitake mycelium or inoculum, and then sealing the filled holes with hot wax. After inoculation, the logs are placed under the closed canopy of a coniferous stand and are left to incubate for 12 to 15 months. Once incubation is complete, the logs are soaked in water for 24 hours. 7 to 10 days after soaking, shiitake mushrooms will begin to fruit and can be harvested once fully ripe.

Indoor Trays

Indoor growing provides the ability to tightly regulate light, temperature and humidity while excluding contaminants and pests. This allows consistent production, regulated by spawning cycles. This is typically accomplished in windowless, purpose-built buildings, for large scale commercial production.

Indoor tray growing is the most common commercial technique, followed by containerized growing. The tray technique provides the advantages of scalability and easier harvesting. Unlike wild harvests, indoor techniques provide tight control over growing substrate composition and growing conditions. Indoor harvests are much more predictable.

Complete sterilization is not always required or performed during composting. In some cases, a pasteurization step is not included to allow some beneficial microorganisms to remain in the growth substrate.

Specific time spans and temperatures required during stages 3–6 will vary respective to species and variety. Substrate composition and the geometry of growth substrate will also affect the ideal times and temperatures. Pinning is the trickiest part for a mushroom grower, since a combination of carbon dioxide (CO_2) concentration, temperature, light, and humidity triggers mushrooms towards fruiting. Up until the point when rhizomorphs or mushroom "pins" appear, the mycelium is an amorphous mass spread throughout the growth substrate, unrecognizable as a mushroom.

Carbon dioxide concentration becomes elevated during the vegetative growth phase, when mycelium is sealed in a gas-resistant plastic barrier or bag which traps gases produced by the growing mycelium. To induce pinning, this barrier is opened or ruptured. CO_2 concentration then decreases from about 0.08% to 0.04%, the ambient atmospheric level.

Substrates

Mushroom production converts the raw natural ingredients into mushroom tissue, most notably the carbohydrate chitin.

An ideal substrate will contain enough nitrogen and carbohydrate for rapid mushroom growth. Common bulk substrates include:

- Wood chips or sawdust
- Mulched straw (usually wheat, but also rice and other straws)
- Strawbedded horse or poultry manure
- Corncobs
- Waste or recycled paper
- coffee pulp or grounds
- Nut and seed hulls
- Cottonseed hulls
- Cocoa bean hulls
- Cottonseed meal

- Soybean meal
- Brewer's grain
- Ammonium nitrate
- Urea

Mushrooms metabolize complex carbohydrates in their substrate into glucose, which is then transported through the mycelium as needed for growth and energy. While it is used as a main energy source, its concentration in the growth medium should not exceed 2%. For ideal fruiting, closer to 1% is ideal.

Pests and Diseases

Parasitic insects, bacteria and other fungi all pose risks to indoor production. The sciarid fly or phorid fly may lay eggs in the growth medium, which hatch into maggots and damage developing mushrooms during all growth stages. Bacterial blotch caused by *Pseudomonas* bacteria or patches of *Trichoderma* green mold also pose a risks during the fruiting stage. Pesticides and sanitizing agents are available to use against these infestations. Biological controls for insect sciarid and phorid flies have also been proposed.

A recent epidemic of Trichoderma green mold has significantly affected mushroom production: "From 1994-96, crop losses in Pennsylvania ranged from 30 to 100%".

Karela

Momordica charantia often called bitter melon, bitter gourd or bitter squash in English, has many other local names. Goya from the indigenous language of Okinawa and karavella from Sanskrit are also used by English-language speakers. It is a tropical and subtropical vine of the family Cucurbitaceae, widely grown in Asia, Africa, and the Caribbean for its edible fruit, which is extremely bitter. Its many varieties differ substantially in the shape and bitterness of the fruit.

Bitter melon originated on the Indian subcontinent, and was introduced into China in the 14th century.

This herbaceous, tendril-bearing vine grows to 5 m. It bears simple, alternate leaves 4–12 cm across, with three to seven deeply separated lobes. Each plant bears separate yellow male and female flowers. In the Northern Hemisphere, flowering occurs during June to July and fruiting during September to November.

The fruit has a distinct warty exterior and an oblong shape. It is hollow in cross-section, with a relatively thin layer of flesh surrounding a central seed cavity filled with large, flat seeds and pith. The fruit is most often eaten green, or as it is beginning to turn yellow. At this stage, the fruit's flesh is crunchy and watery in texture, similar to cucumber, chayote or green bell pepper, but bitter. The skin is tender and edible. Seeds and pith appear white in unripe fruits; they are not intensely bitter and can be removed before cooking.

As the fruit ripens, the flesh (rind) becomes tougher, more bitter, and too distasteful to eat. On the other hand, the pith becomes sweet and intensely red; it can be eaten uncooked in this state, and is a popular ingredient in some Southeast Asian salads.

When the fruit is fully ripe, it turns orange and mushy, and splits into segments which curl back dramatically to expose seeds covered in bright red pulp.

Some parts of the plant including arils are toxic, causing vomiting, diarrhea, and even death in children.

Varieties

Bitter melon comes in a variety of shapes and sizes. The cultivar common to China is 20–30 cm long, oblong with bluntly tapering ends and pale green in colour, with a gently undulating, warty surface. The bitter melon more typical of India has a narrower shape with pointed ends, and a surface covered with jagged, triangular "teeth" and ridges. It is green to white in colour. Between these two extremes are any number of intermediate forms. Some bear miniature fruit of only 6–10 cm in length, which may be served individually as stuffed vegetables. These miniature fruit are popular in Pakistan (common name 'Karela'), India, Nepal and other countries in South Asia. The sub-continent variety is most popular in Pakistan.

Culinary Uses

Bitter melon is generally consumed cooked in the green or early yellowing stage. The young shoots and leaves of the bitter melon may also be eaten as greens.

Bitter melon is often used in Chinese cooking for its bitter flavour, typically in stir-fries (often with pork and *douchi*), soups, and herbal teas. It has also been used in place of hops as the bittering ingredient in some Chinese and Okinawan beers.

It is very popular throughout South Asia. In North Indian cuisine, it is often prepared with potatoes and served with yogurt on the side to offset the bitterness, used in *sabzi* or stuffed with spices and then cooked in oil. In Southern India, it is used in the dishes *thoran/thuvaran* (mixed with grated coconut), *mezhukkupuratti* (stir fried with spices), *theeyal* (cooked with roasted coconut) and *pachadi* (which is considered a medicinal food for diabetics). Other popular recipes include preparations with curry, deep fried with peanuts or other ground nuts, and *Pachi Pulusu*, a soup with fried onions and other spicesIn Konkan region of Maharashtra, salt is added to finely chopped bitter gourd and then it is squeezed, removing its bitter juice to some extent. After frying this with different spices, less bitter and crispy preparation is served with grated coconut.

In Pakistan and Bangladesh, bitter melon is often cooked with onions, red chili powder, turmeric powder, salt, coriander powder, and a pinch of cumin seeds. Another dish in Pakistan calls for whole, unpeeled bitter melon to be boiled and then stuffed with cooked minced beef, served with either hot *tandoori* bread, *naan*, *chappati*, or with *khichri* (a mixture of lentils and rice).

Bitter melon is a significant ingredient in Okinawan cuisine, and is increasingly used in mainland Japan. It is popularly credited with Okinawan life expectancies being higher than the already long Japanese ones.

In Indonesia, bitter melon is prepared in various dishes, such as *gado-gado*, and also stir fried, cooked in coconut milk, or steamed.

In Vietnam, raw bitter melon slices consumed with dried meat floss and bitter melon soup with shrimp are popular dishes. Bitter melons stuffed with ground pork are served as a popular summer soup in the south. It is also used as the main ingredient of "stewed bitter melon". This dish is usually cooked for the T¿t holiday, where its "bitter" name is taken as a reminder of the poor living conditions experienced in the past.

In the Philippines, bitter melon may be stir-fried with ground beef and oyster sauce, or with eggs and diced tomato. The dish *pinakbet*, popular in the Ilocos region of Luzon, consists mainly of bitter melons, eggplant, okra, string beans, tomatoes, lima beans, and other various regional vegetables altogether stewed with a little *bagoong*-based stock.

In Nepal, bitter melon is prepared as a fresh pickle called *achar*. For this, the bitter gourd is cut into cubes or slices and sautéed covered in oil and a sprinkle of water. When it is softened and reduced, it is minced in a mortar with a few cloves of garlic, salt and a red or green pepper. It is also sautéed to golden-brown, stuffed, or as a curry on its own or with potatoes.

In Trinidad and Tobago bitter melons are usually sauteed with onion, garlic and scotch bonnet pepper until almost crisp.

Summer Squash

Summer squash are a subset of squashes that are harvested when immature, while the rind is still tender and edible. Nearly all summer squash squashes are varieties of *Cucurbita pepo*, though not all *Cucurbita pepo* are considered summer squashes. Most summer squash have a bushy growth habit, unlike the rambling vines of many winter squashes. The name "summer squash" refers to the short storage life of these squashes, unlike that of winter squashes.

Summer squashes include:

- Cousa squash, pale-coloured Zucchini varieties purportedly of Middle Eastern or West Asian descent. Not to be confused with Cushaw, a type of winter squash.
- Pattypan squash (Scallop squash)
- Tromboncino or Zucchetta, unusual among summer squash as being a vining plant and a *Cucurbita moschata* variety.
- Pygon Squash, a summer squash left to over ripen, known for their larger dimensions
- Yellow crookneck squash
- Yellow summer squash
- Zucchini (courgette)
- Immature Ridge gourd luffa is used as a summer squash in India, where it is known as *turai* or *dodka*.

Zucchini

The zucchini or courgette is a summer squash which can reach nearly a metre in length, but which is usually harvested at half that size or less. Along with certain other squashes and pumpkins, it belongs to the species *Cucurbita pepo*. Zucchini can be dark or light green. A related hybrid, the golden zucchini, is a deep yellow or orange colour.

In a culinary context, the zucchini is treated as a vegetable, which means it is usually cooked and presented as a savory dish or accompaniment. Botanically, however, the zucchini is a fruit, being the swollen ovary of the zucchini flower. The zucchini or courgette originates from Italy.

Flower

The female flower is a golden blossom on the end of each emergent zucchini. The male flower grows directly on the stem of the zucchini plant in the leaf axils (where leaf petiole meets stem), on a long stalk, and is slightly smaller than the female. Both flowers are edible, and are often used to dress a meal or to garnish the cooked fruit.

Firm and fresh blossoms that are only slightly open are cooked to be eaten, with pistils removed from female flowers, and stamens removed from male flowers. The stem on the flowers can be retained as a way of giving the cook something to hold onto during cooking, rather than injuring the delicate petals, or they can be removed prior to cooking, or prior to serving. There are a variety of recipes in which the flowers may be deep fried as fritters or tempura (after dipping in a light tempura batter), stuffed, sautéed, baked, or used in soups.

History and Etymology

Zucchini, like all squash, has its ancestry in the Americas. However, the varieties of squash typically called "zucchini" were developed in Italy, many generations after their introduction from the Americas.

In all probability, this occurred in the very late 19th century, probably near Milan; early varieties usually included the names of nearby cities in their names. The alternative name *courgette* is from the French word for the vegetable, with the same spelling, and is commonly used in France, The Netherlands, Ireland, and the United Kingdom. It is a diminutive of *courge*, French for squash. "Zucca" is the Italian word for pumpkin/ squash and "zucchina" (zucca + ina = little) is its diminutive, becoming "zucchine" in the plural. "zucchino", the masculine form, ("zucchini" in the plural), is commonly used in the dialect of Tuscany as the name of the fruit and sometimes is improperly used in Italian as the name of the plant. An Italian dictionary called "lo Zingarelli 1995, Zanichelli editor", gives both forms, but some others, like "Devoto-Oli, Le Monnier editor", but "Treccani, Treccani editor", the highest linguistic authority, gives 'zucchina' as the correct Italian word, and the masuline form as regional or vernacular. "Zucchini" is used in Tuscany, Australia, Canada and the United States. 'Zucchini', the plural in the dialect of Tuscany, is one of the plural forms in English (along with 'zucchinis') as well as the singular form. The first records of zucchini in the United States date to the early 1920s. It was almost certainly brought over by Italian immigrants and probably was first cultivated in the United States in California. In South Africa, they are called *baby marrow*.

Culinary Uses

When used for food, zucchini are usually picked when under 20 cm (8 in.) in length, when the seeds are still soft and immature. Mature zucchini can be as large as a baseball bat. The larger ones are often fibrous. A zucchini with the flowers attached are a sign of a truly fresh and immature fruit, and are especially sought by many people for its sweeter flavour.

Unlike cucumber, zucchini is usually served cooked. It can be prepared using a variety of cooking techniques, including steamed, boiled, grilled, stuffed and baked, barbecued, fried, or incorporated in other recipes such as *soufflés*. Zucchini can also be baked into bread similar to banana bread or incorporated into a cake mix. Its flowers can be eaten stuffed and are a delicacy when deep fried, as *tempura*.

***Figure:** Grilled zucchini*

The zucchini has a delicate flavour and requires little more than quick cooking with butter or olive oil, with or without fresh herbs. The skin is left in place. Quick cooking of barely wet zucchini in oil or butter allows the fruit to partially boil and steam, with the juices concentrated in the final moments of frying when the water has gone, prior to serving. Zucchini can also be eaten raw, sliced or shredded in a cold salad, as well as lightly cooked in hot salads, as in Thai or Vietnamese recipes. Mature (larger sized) zucchini are well suited for cooking in breads.

In 2005, a poll of 2,000 people revealed the zucchini (courgette) to be Britain's 10th favourite culinary vegetable.

In Mexico, the flower (known as *flor de calabaza*) is often cooked in soups or used as a filling for quesadillas. The fruit is used in stews, soups (i.e. caldo de res, de pollo or de pescado, mole de olla, etc.) and other preparations. Both the flower and the fruit are eaten readily throughout Mexico.

In Italy, zucchini are served in a variety of ways, in the pan, in the oven, boiled or fried, alone or in combination with other ingredients.

At home and in some restaurants, it's possible to eat the flowers as well, cooked in deep-frying, known as *fiori di zucca.*

In France zucchini are a key ingredient in *ratatouille*, a stew of summer fruits and vegetables prepared in olive oil and cooked for an extended time over low heat. The dish, originating near present-day Nice, is served as a side dish or on its own at lunch with bread. Zucchini are stuffed with meat with other fruits like tomatoes or bell peppers in a dish named *courgette farcie* (stuffed zucchinis).

In Turkey, zucchini is the main ingredient in the popular dish *mücver*, or "zucchini pancakes", made from shredded zucchini, flour and eggs, lightly fried in olive oil and eaten with yogurt.They are also used not infrequently in kebabs along with various meats.

In the Levant, zucchini is stuffed with minced meat and rice plus herbs and spices and steamed. It is also used in various kinds of stew. Stews that have low salinity are favourable in such cooking. It can also be stuffed with a mixture of rice, meat and eaten with yogurt.

In Greece, zucchini is usually fried or boiled with other fruits (often green chili peppers and eggplants). It is served as an *hors d'œuvre* or as a main dish, especially during fasting seasons. Zucchini is also often stuffed with minced meat, rice and herbs and served with avgolemono sauce. In several parts of Greece, the flowers of the plant are stuffed with white cheese, usually feta or mizithra cheese, or with a mixture of rice, herbs and occasionally minced meat. Then they are deep-fried or baked with tomato sauce in the oven.

In Bulgaria, zucchini are fried and then served with a dip, made from yogurt, garlic and dill. Another popular dish is oven-baked zucchini—sliced or grated—covered with a mixture of eggs, yogurt, flour and dill.

In Egypt, zucchini are cooked with tomato sauce, garlic, and onions.

Nutrition

The zucchini fruit is low in calories (approximately 15 food calories per 100 g fresh zucchini) and contains useful amounts of folate (24 μg/ 100 g), potassium (280 mg/100 g) and vitamin A (384 IU [115 μg]/100 g).

Cultivation

Zucchini is one of the easiest vegetables to cultivate in temperate climates. As such, it has a reputation among home gardeners for overwhelming production. One good way to control over-abundance is to harvest the flowers, which are an expensive delicacy in markets because of the difficulty in storing and transporting them. The male flower is borne on the end of a stalk and is longer lived.

While easy to grow, zucchini, like all squash, requires plentiful bees for pollination. In areas of pollinator decline or high pesticide use, such as mosquito-spray districts, gardeners often experience fruit abortion, where the fruit begins to grow, then dries or rots. This is due to an insufficient number of pollen grains delivered to the female flower. It can be corrected by hand pollination or by increasing the bee population.

Closely related to zucchini are Lebanese summer squash or *kusa* (not to be confused with Cushaw), but they often are lighter green or even white. Some seed catalogues do not distinguish them. Various varieties of round zucchinis are grown in different countries under different names, such as "Tondo di Piacenza" in Italy and "Ronde de Nice" in France. In the late 1990s American producers in California cultivated and began marketing round yellow and green zucchini known as "8-ball" squash (the yellow ones are sometimes known as "1-ball" or "gold ball"). White zucchini (summer squash) is sometimes seen as a mutation and can appear on the same plant as its green counterpart.

Tromboncino (squash)

Tromboncino, also known as zucchetta, is a type of squash most often used as a summer squash. While nearly all summer squash are cultivars of *Cucurbita pepo*, tromboncino is a cultivar of *Cucurbita moschata*. Also unique is the vining growth habit, similar to many winter squashes, but unlike most other summer squash. It is more tolerant of some common summer squash pests, including squash vine borer, squash bugs, and powdery mildew, than the more common, bushy, *C. pepo* summer squash cultivars, though it is also slower to grow. Colour is usually pale green, fading to beige upon maturity, and it is picked around one foot long for summer squash. It is an heirloom, originally from Liguria, and remains popular throughout Italy and abroad. Tromboncino squash can be left to mature into a winter squash; such is often compared to a watery butternut squash. If left to ripen, the fruits can grow over three feet in length.

Tromboncino is known by many common names, including: zucchetta rampicante, zucchino rampicante, climbing zucchini, climbing crookneck, trombolino d'albenga, and serpentine squash.

Okra

Okra, known in many English-speaking countries as ladies' fingers, bhindi, bamia, or gumbo, is a flowering plant in the mallow family. It is valued for its edible green seed pods. The geographical origin of okra is disputed, with supporters of Guatemalan, West African, Ethiopian and Bangladeshi origins. The plant is cultivated in tropical, subtropical and warm temperate regions around the world.

Vernacular Names

The name *okra* is most often used in the United States, and Philippines with a variant pronunciation, English Caribbean *okro*. The word *okra* is of Nigerian origin and is cognate with *Íkwurå* in the Igbo language spoken in Nigeria. Okra is often known as "lady's fingers" outside of Africa. In various Bantu languages, okra is called *kingombo* or a variant thereof, and this is the origin of its name in Portuguese (*quiabo*), Spanish (*quimbombó* or *guigambó*), Dutch and French (*gombo*), and also possibly of the name "gumbo", used in parts of the United States and English-speaking Caribbean for either the vegetable or a stew based on it.

In India and Pakistan, and often in the United Kingdom, it is called by its Hindi/Urdu name, *bhindi*, *bhendi*, *bendai* or *bhinda*. In Bangladesh and West Bengal, India it is called *dherosh*. In Tamil Nadu and Kerala, India it is called *vendai kai*. In Andhra Pradesh and Karnataka, India it is called *bende kayi*. In Odisha it is called "Bhendi". In China, it is called *qiu kui*. In the Middle East (in Arabic, Farsi, Hebrew, and Turkish), it is called *bamia* or *bamyeh*. In Bosnia and Herzegovina it is called *bamija*. In the Dominican Republic, it is called *molondron*.

Structure and Physiology

The species is an annual or perennial, growing to 2 m tall. It is related to such species as cotton, cocoa, and hibiscus. The leaves are 10–20 cm long and broad, palmately lobed with 5–7 lobes. The flowers are 4–8 cm in diametre, with five white to yellow petals, often with a red or purple spot at the base of each petal. The fruit is a capsule up to 18 cm long, containing numerous seeds.

Abelmoschus esculentus is cultivated throughout the tropical and warm temperate regions of the world for its fibrous fruits or pods containing round, white seeds. It is among the most heat- and drought-tolerant vegetable species in the world and will tolerate soils with heavy clay and intermittent moisture but frost can damage the pods.

In cultivation, the seeds are soaked overnight prior to planting to a depth of 1–2 cm. Germination occurs between six days (soaked seeds) and three weeks. Seedlings require ample water. The seed pods rapidly become fibrous and woody, and, to be edible, must be harvested within a week of the fruit having been pollinated. The fruits are harvested when immature and eaten as a vegetable.

Origin and Distribution

Okra is an allopolyploid of uncertain parentage (proposed parents include *Abelmoschus ficulneus*, *A. tuberculatus* and a reported “diploid” form of okra). Truly wild (as opposed to naturalised) populations are not known with certainty and the species may be a cultigen.

Figure: *Whole plant*

The geographical origin of okra is disputed, with supporters of South Asian, Ethiopian and West African origins. Supporters of a South Asian origin point to the presence of its proposed parents in that region. Supporters of a West African origin point to the greater diversity of okra in that region.

The Egyptians and Moors of the 12th and 13th centuries used the Arabic word for the plant, *bamya*, suggesting it had come from the east. The plant may have entered southwest Asia across the Red Sea or the Bab-el-Mandeb strait to the Arabian Peninsula, rather than north across the Sahara, or from India. One of the earliest accounts is by a Spanish Moor who visited Egypt in 1216, who described the plant under cultivation by the locals who ate the tender, young pods with meal.

From Arabia, the plant spread around the shores of the Mediterranean Sea and eastward. The plant was introduced to the Americas by ships plying the Atlantic slave trade by 1658, when its presence was recorded in Brazil. It was further documented in Suriname in 1686. In compound farms in the rainforest of southeastern Nigeria (Okafor and Fernandes, 1986), farmers have developed a multi-crop system that provides a diversified and continuous production of food, combining species with different maturity periods such as yams, cassava, cocoyams, bananas, plantain, maize, okra, pumpkin, melon, leafy vegetables and a variety of trees and shrubs, 60 of which provide food products. This ensures a balanced diet but also reduces the need for storage in an area where post-harvest losses are high.

Okra may have been introduced to southeastern North America in the early 18th century. It was being grown as far north as Philadelphia by 1748. Thomas Jefferson noted it was well established in Virginia by 1781. It was commonplace throughout the Southern United States by 1800, and the first mention of different cultivars was in 1806.

Nutrition

Okra is a popular health food due to its high fibre, vitamin C, and folate content. Okra is also known for being high in antioxidants. Okra is also a good source of calcium and potassium.

Okra Seed Oil

Greenish-yellow edible okra oil is pressed from okra seeds; it has a pleasant taste and odour, and is high in unsaturated fats such as oleic acid and linoleic acid. The oil content of some varieties of the seed can be quite high, about 40%. Oil yields from okra crops are also high. At 794 kg/ha, the yield was exceeded only by that of sunflower oil in one trial. A 1920 study found that a sample contained 15% oil. A 2009 study found okra oil suitable for use as a biofuel.

Culinary Uses

The products of the plant are mucilaginous, resulting in the characteristic "goo" or slime when the seed pods are cooked; the mucilage contains a usable form of soluble fibre. Some people cook okra this way, others prefer to minimize sliminess; keeping the pods intact, and brief cooking, for example stir-frying, help to achieve this. Cooking with acidic ingredients such as a few drops of lemon juice, tomatoes, or vinegar may help. Alternatively, the pods can be sliced thinly and cooked for a long time so the mucilage dissolves, as in gumbo. The immature pods may also be pickled. Okra is considered a delicacy in the American Deep South, particularly when breaded and deep fried. Several cafes and nationwide eating establishments serve deep fried okra, typically alongside a side order of a sauce such as buttermilk dressing.

In Syria, Tunisia, Egypt, Albania, Bosnia, Greece, Bulgaria, Republic of Macedonia, Iran, Iraq, Jordan, Lebanon, Turkey, and Yemen, and other parts of the eastern Mediterranean, including Palestine, Cyprus and Israel, okra is widely used in a thick stew made with vegetables and meat. In Bosnia and most of West Asia, okra is known as *bamia* or *bamya*. West Asian cuisine usually uses young okra pods, usually cooked whole. In India, the harvesting is done at a later stage, when the pods and seeds are larger.

It is popular in Indian and Pakistani cuisine, where chopped pieces are stir-fried with spices, pickled, salted or added to gravy-based preparations such as *bhindi ghosht* and *sambar*. It is also simmered in coconut based curries or tossed with ground mustard seeds. In India, it is also used in curries. In curries, okra is used whole, trimmed only of excess stalk and keeping the hard conical top, which is discarded at the time of eating. In South India, Okra is cut into small circular pieces about 1/4 inch thick and stick fried in oil with salt and hot pepper powder to make delicious curry. However, when used in sambar it is cut into pieces which are 1 inch thick to prevent it from dissolving when the sambar is let to simmer.

In Malaysia okra is commonly a part of *yong tau foo* cuisine, typically stuffed with processed fish paste (surimi) and boiled with a selection of vegetables and tofu, and served in a soup with noodles.

In Malawi it is preferred cooked and stirred with sodium bicarbonate to make it more slimy. It is then commonly eaten with nsima (pap) made from raw maize flour or maize husks flour.

In the Caribbean islands, okra is eaten in soup. In Curaçao the soup is known as *jambo* which primarily is made out of the okra's mucilage. It is often prepared with fish and *funchi*, a dish made out of cornmeal and boiling water. In Haiti, it is cooked with rice and maize, and also used as a sauce for meat. In Cuba, it is called *quimbombó*, along with a stew using okra as its primary ingredient. In Dominican Republic is eaten as if in salad and also cooked with rice.

It became a popular vegetable in Japanese cuisine towards the end of the 19th century, served with soy sauce and *katsuobushi*, or as tempura.

In the Philippines, okra can be found among traditional dishes like pinakbet, dinengdeng, and sinigang. Because of its mild taste and ubiquity, okra can also be cooked adobo-style, or served steamed or boiled in a salad with tomatoes, onion and bagoong.

Okra forms part of several regional "signature" dishes. *Frango com quiabo* (chicken with okra) is a Brazilian dish especially famous in the region of Minas Gerais, and it is the main ingredient of caruru, a bahian food with dende oil. Gumbo, a hearty stew whose key ingredient is okra, is found throughout the Gulf Coast of the United States and in the South Carolina Lowcountry. Deep- or shallow-fried okra coated with cornmeal, flour, etc. is widely eaten in the southern United States. Okra is also an ingredient expected in callaloo, a Caribbean dish and the national dish of Trinidad and Tobago. It is also a part of the national dish of Barbados *coucou* (turned cornmeal). Okra is also eaten in Nigeria, where draw soup is a popular dish, often eaten with *garri* or cassava. In Vietnam, okra is the important ingredient in the dish *canh chua*. Okra slices can also be added to *ratatouille*.

Figure: *A variety of okra pods with a dark red pigmentation*

Okra leaves may be cooked in a similar way to the greens of beets or dandelions. Since the entire plant is edible, the leaves are also eaten raw in salads. Okra seeds may be roasted and ground to form a caffeine-free substitute for coffee. When importation of coffee was disrupted by the American Civil War in 1861, the *Austin State Gazette* said "An acre of okra will produce seed enough to furnish a plantation of fifty negroes with coffee in every way equal to that imported from Rio."

Alternative Medicinal Properties

Naturopathic medicines include okra for intestinal and irritable bowel dysfunction due to its mucilaginous properties. Utilizing the whole pod during cooking maintains the potency and the effectiveness of the pod contents. These are still present when chopped, but to a lesser degree. The consistency of the mucilaginous seeds are a good binder in the recipe in which they are cooked too, giving dishes a nice glossy appearance. In complementary medicine it is believed that the gut and intestines are the main/dominant organs of the body and that all foods which facilitate the digestive process and binary systems are beneficial to the body as a whole, for this reason, providing a natural lining to the intestinal mucosa, prevents leakages from ailments such as ulcerations, bacterial imbalances and general dyspepsia. Thus okra's fibrous and mucilaginous properties are likely to have excellent results in lowering inflammation, increasing active transportation of nutritional conversion (take up) and reduce fluid retention and subsequent secondary complications associated with the latter.

Unspecified parts of the plant were reported in 1898 to possess diuretic properties; this is referenced in numerous sources associated with herbal and traditional medicine.

Some studies are being developed targeting okra extract as remedy to manage diabetes.

Common Diseases

The most common disease afflicting the okra plant is verticillium wilt, often causing a yellowing and wilting of the leaves. Other diseases include powdery mildew in dry tropical regions, leaf spots, and root-knot nematodes.

Oxalic Acid

Oxalic acid is a naturally occurring colourless organic acid found in many plants including okra. Okra contains 0.05g/100g of oxalic acid. The toxicity of oxalic acid is due to kidney failure, which arises because it causes precipitation of solid calcium oxalate, the main component of kidney stones. Oxalic acid can also cause joint pain due to the formation of similar precipitates in the joints.

Sweet Potatoes

From creamy white sweet potatoes to nutritious purple tubers, there are sweet potato varieties tailored to every gardener's region and palate. Learn how to grow, harvest, cook and store the variety of your choice.

Eggplant

Eggplant (*Solanum melongena*) is a species of nightshade commonly known in British English as aubergine and also known as melongene, garden egg, or guinea squash. It is known in South Asia, Southeast Asia and South Africa as brinjal. It bears a fruit of the same name (commonly either "eggplant" in American and Australian English or "aubergine" in British English) that is widely used in cooking, most notably as an important ingredient in dishes such as moussaka and ratatouille. As a member of the genus *Solanum*, it is related to both the tomato and the potato. It was originally domesticated in India and Bangladesh from the wild nightshade, the thorn or bitter apple, *S. incanum*.

The eggplant is a delicate, tropical perennial often cultivated as a tender or half-hardy annual in temperate climates. It grows 40 to 150 cm (16 to 57 in) tall, with large, coarsely lobed leaves that are 10 to 20 cm (4–8 in) long and 5 to 10 cm (2–4 in) broad. Semiwild types can grow much larger, to 225 cm (7 ft) with large leaves over 30 cm (12 in) long and 15 cm (6 in) broad. The stem is often spiny. The flower is white to purple, with a five-lobed corolla and yellow stamens. The egg-shaped glossy black fruit has white flesh with a meaty texture. The cut surface of the flesh rapidly turns brown when the fruit is cut open. On wild plants, the fruit is less than 3 cm (1.2 in) in diametre, but very much larger in cultivated forms, reaching 30 cm (12 in) or more in length.

The fruit is botanically classified as a berry and contains numerous small, soft seeds which are edible, but have a bitter taste because they contain nicotinoid alkaloids (being a relative of tobacco).

Names and Etymology

Figure: *Closeup of an eggplant flower of a long-fruited Chinese variety in Hong Kong.*

Some 18th-century European cultivars were yellow or white and resembled goose or hen's eggs, hence the name "eggplant."

The Arabic name is the common source of almost all European names for this plant, but through two distinct paths of transmission, with the *melongene* family coming through the eastern Mediterranean, and the *aubergine* family through the western Mediterranean.

In the eastern Mediterranean, Byzantine Greek borrowed *bâinjân* as ìåëéôæÜíá *melitzána*, influenced by Greek 'black'. That form came into medieval Latin as *melongena*, which was used in the botanical works of Tournefort and Linnaeus. Though *melongene* has become obsolete in the standard English, as has the French *melanjan*, it persists in the Caribbean English *melongene* or *meloongen*. The usual word in Italian remains *melanzana*.

Even the archaic English name *mad-apple* comes from the *melongena* family: in Italian, the word *melanzana* was reinterpreted in Italian as *mela insana*, and translated into English as *mad apple*.

In the western Mediterranean, *(al)-bâinjân* became Spanish *berenjena*, Catalan as *albergínia*, and Portuguese *beringela*. The Catalan form was borrowed by French as *aubergine*, which was then borrowed into British English.

In Eastern Slavic languages, such as Russian and Ukrainian, the word *baklazhan* is used, while Turkish has *patlýcan*.

In South Asian, South African, Malaysian, Singaporean, and West Indian English, the fruit is called *brinjal*, from the Portuguese. The Indic name *baingan* or *baigan* is also sometimes used in South Asian English.

In Kiswahili, it is called *biringanya*.

History

The plant is native to the Indian Subcontinent. It has been cultivated in southern and eastern Asia since prehistory. The first known written record of the plant is found in *Qí mín yào shù*, an ancient Chinese agricultural treatise completed in 544. The numerous Arabic and North African names for it, along with the lack of the ancient Greek and Roman names, indicate it was introduced throughout the Mediterranean area by the Arabs in the early Middle Ages. A book on agriculture by Ibn Al-Awwam in 12th century Arabic Spain described how to grow aubergines. There are records from later medieval Catalan and Spanish.

The aubergine is unrecorded in England until the 16th century. An English botany book in 1597 stated:

This plant groweth in Egypt almost everywhere... bringing forth fruit of the bigness of a great cucumber.... We have had the same in our London gardens, where it hath borne flowers, but the winter approaching before the time of ripening, it perished: nothwithstanding it came to bear fruit of the bigness of a goose egg one extraordinary temperate year... but never to the full ripeness.

Because of the plant's relationship with the Solanaceae (nightshade) family, the fruit was at one time believed to be extremely poisonous. The flowers and leaves can be poisonous if consumed in large quantities due to the presence of solanine.

Cultivated Varieties

Different varieties of the plant produce fruit of different size, shape, and colour, though typically purple. The most widely cultivated varieties (cultivars) in Europe and North America today are elongated ovoid, 12–25 cm long (4½ to 9 in) and 6–9 cm broad (2 to 4 in) in a dark purple skin.

A much wider range of shapes, sizes and colours is grown in India and elsewhere in Asia. Larger varieties weighing up to a kilogram (2.2

pounds) grow in the region between the Ganges and Yamuna rivers, while smaller varieties are found elsewhere. Colours vary from white to yellow or green, as well as reddish-purple and dark purple. Some cultivars have a colour gradient, from white at the stem to bright pink to deep purple or even black. Green or purple cultivars in white striping also exist. Chinese varieties are commonly shaped like a narrower, slightly pendulous cucumber, and are sometimes called Japanese eggplants in North America.

Oval or elongated oval-shaped and black-skinned cultivars include 'Harris Special Hibush', 'Burpee Hybrid', 'Black Magic', 'Classic', 'Dusky', and 'Black Beauty'. Slim cultivars in purple-black skin include 'Little Fingers', 'Ichiban', 'Pingtung Long', and 'Tycoon'; in green skin, 'Louisiana Long Green' and 'Thai (Long) Green'; in white skin, 'Dourga'. Traditional, white-skinned, egg-shaped cultivars include 'Casper' and 'Easter Egg'. Bicoloured cultivars with colour gradient include 'Rosa Bianca', 'Violetta di Firenze', 'Bianca Smufata di Rosa' (heirloom), and 'Prosperosa' (heirloom). Bicoloured cultivars with striping include 'Listada de Gandia' and 'Udumalapet'. In some parts of India, miniature varieties (most commonly called *vengan*) are popular. A particular variety of green brinjal known as *Matti gulla* is grown in Matti, a village of the Udupi district in Karnataka state.

Cultivation

In tropical and subtropical climates, eggplant can be sown directly into the garden. Eggplant grown in temperate climates fares better when transplanted into the garden after all danger of frost is passed. Seeds are typically started eight to 10 weeks prior to the anticipated frost-free date. Many pests and diseases which afflict other solanaceous plants, such as tomato, pepper (capsicum), and potato, are also troublesome to eggplants. For this reason, it should not be planted in areas previously occupied by its close relatives. Four years should separate successive crops of eggplants. Common North American pests include the potato beetles, flea beetles, aphids, and spider mites. (Adults can be removed by hand, though flea beetles can be especially difficult to control.) Good sanitation and crop rotation practices are extremely important for controlling fungal disease, the most serious of which is *Verticillium*.

Spacing should be 45 cm (18 in.) to 60 cm (24 in.) between plants, depending on cultivar, and 60 cm to 90 cm (24 to 36 in.) between rows, depending on the type of cultivation equipment being used. Mulching will help conserve moisture and prevent weeds and fungal diseases. The flowers are relatively unattractive to bees and the first blossoms often do not set fruit. Hand pollination will improve the set of the first blossoms. Fruits are typically cut from the vine just above the calyx owing to the somewhat woody stems. Flowers are complete, containing both female and male structures, and may be self-pollinated or cross-pollinated.

Crop Protection

Shoot and fruit borer: *Leucinodes orbonalis*

Symptoms of damage

- Withering of terminal shoots/ dead hearts
- Bore holes on shoots and fruits plugged with excreta
- Shedding of flower buds
- Withering and drying of leaves

Identification of pest

- Eggs - Creamy white eggs
- Larva - pink in colour
- Pupa- greyish boat shaped cocoon
- Adult- a medium sized moth
- Forewings- having black and brown patches and dots on white colour
- Hind wings – opalescent with black dots

Management

- Remove the affected terminal shoot showing boreholes.
- Remove the affected fruits and destroy.
- Avoid continuous cropping of brinjal crop
- Grow the varieties with long and narrow fruits in endemic areas
- Install pheromone trap@12/ha
- Encourage the activity of larval parasitoids:
    - o *Pristomerus testaceus*
    - o *Cremastus flavoorbitalis*
- Avoid use of synthetic pyrethroids
- Avoid using insecticides at the time of fruit maturation and harvest
- Neem seed kernel extract (NSKE) 5 % or any one of the following chemicals starting from one month after planting at 15 days interval

Insecticide	***Dose***
Azadirachtin 1.0% EC (10000 ppm)	3.0 ml/lit.
Azadirachtin 0.03 % WSP (300 ppm)	5.0 g/lit.
Chlorpyrifos 20 % EC	1.0 ml/lit.
Dimethoate 30 % EC	7.0 ml/10 lit.
Emamectin benzoate 5 % SG	4 g/10 lit.
Flubendiamide 20 WDG	7.5 g/10 lit.
Phosalone 35 % EC	1.5 ml/lit.
Quinalphos 25 % EC	1.5 ml/lit.
Thiodicarb 75 % WP	2.0 g/lit.
Thiometon 25 % EC	1.0 ml/lit.
Trichlorofon 50 % EC	1.0 ml/lit.
Triazophos 40 % EC	2.5 ml/lit.

Stem borer: Euzophera perticella

Symptoms of damage

- Top shoots of young plants droop and wither.
- Older plants become stunted.
- Fruit bearing is affected

Identification of pest

- Egg- Cream, scale-like
- Larva - Fully grown larva is creamy white
- Adult - Greyish brown,
- Forewings- with transverse line and
- Hind wings - white in colour

Management

- Collect and destroy the damaged and dead plants
- Light trap @ 1/ha to attract and kill adults
- Spray neem oil 2ml/lit
- Avoid using synthetic pyrethriods causing resurgence

Bud Worm: Scrobipalpa Blapsigona

Symptoms of damage:

- Shriveling and shedding of flower buds

Identification of Pest

- Larvae - pale whitish with pink tinge
- Adult - small with fringed wings.

Management

- Spray neem oil 2ml/lit

Hadda/spotted Beetle: H.vigintioito Punctata

Symptoms of damage

- Scrapping of chlorophyll
- Skeletonization and drying of leaves

Identification of Pest

- Eggs- Cigar shaped, yellow in colour
- Grub: Yellowish bearing six rows of longitudinal spines.
- Pupa: Yellowish with spines on posterior part and anterior portion being devoid of spines.
- Adult: *H. vigintioctopunctata* : 14 spots on each elytra, deep red.

Management

- Collect damaged leaves with grubs and egg masses and destroy them
- Shake plants to dislodge grubs, pupae and adults and destroy
- Conserve natural enemies in brinjal ecosystem
- Spray any one of the insecticides
    - o Carbaryl 50 WP @ 3g/lit

Ash weevils: Myllocerus subfasciatus, M. discolour, M. viridanus

Symptoms of damage

- Notching of leaf margins
- Grubs feed on roots - wilting of plants

Identification of pest

- Larva: grub - small and apodous
- Pupa: Pupates in soil in earthern cocoons
- Adult : *M. maculosus:* greenish white with dark lines on elytra
- *M. subfasciatus*: Brown
- *M. discolour*: Brown and white spots
- *M. viridanus*: Small light green weevil

Management

- Collect and destroy adults
- Apply Neem cake @ 500 kg / ha at the time of last ploughing

- In endemic areas, apply carbofuran 3 G @ 15 kg/ ha – 15 days after planting
- Spray carbaryl 50 WP @ 3g + wettable su

Brown Leaf Hopper: Cestius phycitis

Symptoms of damage:

- Reduction in size of leaves
- Shortened petioles,
- Excessive growth of branches general stunting of plants,
- Conversion of floral parts into leafy structures and
- Plants become bushy
- Fruiting is rare.
- Vector of little leaf of brinjal

Identification of Pest

- Small light brown leaf hopper

Management

- Rogue out infected plants destroy them.
- Before transplantation dip the seedlings in 0.2% carbofuran 50 STD solution(control insect vectors).
- Spray with methyl parathion 0.5% or dimethoate 0.3%.

Lace Wing Bug: Urentius Hystricellus

Symptoms of damage:

- Yellowing of leaves
- Affected leaves covered with exuviae and excreta

Identification of Pest

- Egg -white nibble shaped eggs
- Nymph – yellowish white with prominent spines
- Adult
- Dorsal side - straw coloured
- Ventral side – black coloured
- Pronotum and forewings reticulated

Management

- Spray dimethoate 30 EC @ 1 lit / ha or methyl demeton 25 EC @ 1lit /ha

Statistics

According to FAO in 2012, production of eggplant is highly concentrated, with 90% of output coming from five countries. China is the top producer (58% of world output) and India is second (25%), followed by Iran, Egypt and Turkey. More than 4,000,000 acres (1,600,000 ha) are devoted to the cultivation of eggplant in the world.

Varieties

- *Solanum melongena* var. *esculentum* common eggplant, including white varieties, with many cultivars
- *Solanum melongena* var. *depressum* dwarf eggplant
- *Solanum melongena* var. *serpentium* snake eggplant

Genetically Engineered Variety

Bt brinjal is a transgenic eggplant which has a gene from the soil bacterium *Bacillus thuringiensis* inserted into it. This variety was designed to give the plant resistance against lepidopteran insects like the brinjal fruit and shoot borer (*Leucinodes orbonalis*) and fruit borer (*Helicoverpa armigera*).

On 9 February 2010, the Indian Environment Minister, Jairam Ramesh, imposed a moratorium on the cultivation of Bt brinjal. His decision was made after protest from several groups responding to regulatory approval of the cultivation of Bt brinjal in October, 2009. Ramesh stated the moratorium will last "for as long as it is needed to establish public trust and confidence".

Synonyms

- The eggplant is quite often featured in the older scientific literature under the junior synonyms *S. ovigerum* and *S. trongum*.

Figure: *Segmented purple eggplant.*

The eggplant has a long history of taxonomic confusion with the scarlet and Ethiopian eggplants, known as *gilo* and *nakati*, and described by Linnaeus as *S. aethiopicum*. The eggplant was sometimes considered a variety *violaceum* of that species. *S. violaceum* of de Candolle applies to Linnaeus' *S. aethiopicum*. There is an actual *S. violaceum*, an unrelated plant described by Ortega, which used to include Dunal's *S. amblymerum* and was often confused with the same author's *S. brownii*.

Like the potato and *Solanum lichtensteinii*, but unlike the tomato, which then was generally put in a different genus, the eggplant was also described as *S. esculentum*, in this case once more in the course of Dunal's work. He also recognized varieties *aculeatum*, *inerme* and *subinerme* at that time. Similarly, H.C.F. Schuhmacher and Peter Thonning named the eggplant as *S. edule*, which is also a junior synonym of sticky nightshade (*S. sisymbriifolium*). Scopoli's *S. zeylanicum* refers to the eggplant, and that of Blanco to *S. lasiocarpum*.

Bt Brinjal

The Bt brinjal is a suite of transgenic brinjals (also known as an eggplant or aubergine) created by inserting a crystal protein gene (*Cry1Ac*) from the soil bacterium *Bacillus thuringiensis* into the genome of various brinjal cultivars. The insertion of the gene, along with other genetic elements such as promoters, terminators and an antibiotic resistance marker gene into the brinjal plant is accomplished using *Agrobacterium*-mediated genetic transformation. The Bt brinjal has been developed to give resistance against lepidopteron insects, in particular the Brinjal Fruit and Shoot Borer (*Leucinodes orbonalis*)(FSB). Mahyco, an Indian seed company based in Jalna, Maharashtra, has developed the Bt brinjal.

The genetically modified brinjal event is termed Event EE 1 and Mahyco have also applied for approval of two brinjal hybrids. The Event EE 1 was introgressed by plant breeding into various local varieties by University of Agricultural Sciences, Dharwad and Tamil Nadu Agricultural University, Coimbatore. Some of the cultivars of brinjal include: Malpur local, Manjari gota, Kudachi local, Udupi local, 112 GO, and Pabkavi local. It was approved for commercialization in India in 2009, but - after an apparent public outcry and rounds of debates in which representatives from Mahyco, the scientific community, and NGO's spoke on the topic - the Indian Environment Minister, Jairam Ramesh, facilitated a moratorium on its release until further, unspecified, tests were conducted. Bt brinjal was approved for commercial release in Bangladesh in 2013.

Mahyco licensed and used the *cry1Ac* gene obtained from Monsanto and two supporting genes (*nptII* and *aad*). The *cry1Ac* gene is under the transcriptional control of an enhanced cauliflower mosaic virus 35S (CaMV35S) promoter, which ensures the gene is expressed in all the brinjals tissue throughout its complete life cycle. *NptII* and *aad* are selectable marker genes, *nptII* is used to identify transgenic plants from non-transgenic, and *aad* is used to identify the transformed bacteria used during the development of the construct. *Aad* contains a bacterial promoter and is not expressed in the Bt brinjal. The completed construct was inserted into young cotyledons from the brinjal plants using an *Agrobacterium*-mediated technique. *Agrobacterium* naturally inserts DNA into plants from its Ti plasmid, and scientists use this to insert genes of interest into various plants. The transformed plants were regenerated and analyzed for the presence of the gene through Southern blotting. The plants' progeny were also analyzed to identify lines segregating in a Mendelian fashion.

Effective against Pests

When fruit and shoot borer larvae feed on Bt brinjal plants, they ingest the Bt protein Cry1Ac along with plant tissue. In the insect gut, which is alkaline with a pH >9.5, the protein is solubilized and activated by gut proteases. The Bt protein binds to specific receptor proteins present in the insect membrane, resulting in pore formation in the membranes. This leads to disruption of digestive processes, paralysis, and subsequent death of the fruit and shoot borer larvae.

Attempted Commercialization in India

The first agreement to develop Bt Brinjal was signed in 2005 between India's leading seed company-Maharashtra Hybrid Seeds Company, better known as Mahyco (a subsidiary of Monsanto), and two agricultural universities-University of Agricultural Sciences (UAS) in Dharwad and Tamil Nadu Agricultural University (TNAU) in Coimbatore.

An expert committee (EC-I) was set up in 2006 to examine the biosafety data presented by Mahyco. They concluded that while the current data demonstrated that Bt brinjal was safe and equivalent to its non-Bt counterpart, more studies were required to re-affirm the findings and further trials were needed to ascertain the benefits from Bt brinjal with respect to existing methods for pest management and pesticide reduction. They recommended that large scale trials be allowed to go ahead. In 2009, a second expert committee (EC-II) examined the data from these trials. They concluded that adequate safety tests had been performed, stating that "the benefits of Bt brinjal event EE-I developed by M/s Mahyco far outweigh the perceived and projected risks", and advised the Genetic Engineering Appraisal Committee (GEAC) to recommend commercialization of the Bt brinjal.

The GEAC cleared Bt brinjal for commercialization on 14 October 2009. Following concerns raised by some scientists, farmers and anti-GM activists, the government of India officially announced on 9 February 2010 that it needed more time before releasing Bt brinjal, with Indian Environment Minister Jairam Ramesh saying that there is no overriding urgency to introduce Bt brinjal in India. On 17 February 2010, Jairam Ramesh reiterated that the Centre had only imposed a moratorium on the release of transgenic brinjal hybrid, and not a permanent ban, saying that "until we arrive at a political, scientific and societal consensus, this moratorium will remain". Companies with any seeds of Bt brinjal will have to register the details with the government, and the National Bureau of Plant Genetic Resources (NBPGR) was made responsible for storage of all the Bt brinjal seeds in India. Independent testing labs are currently being set up.

An irregularity was also brought to the notice of the Karnataka Biodiversity Board by Environment Support Group, a charitable trust in Bengaluru, in February 2010. It found that agencies accessed at least 10 brinjal varieties from Karnataka and Tamil Nadu without seeking prior consent of the National Biodiversity Authority and state biodiversity boards. Mahyco became India's first commercial entity to be accused of bio-piracy, or misappropriation, of local germplasm.

Fennel

Fennel is both a vegetable and an herb, depending on which variety you grow. Growing crunchy bulb fennel (also called finocchio) is easy in spring and fall, or you can keep a feathery mound of perennial fennel as a steady source of fennel fronds.

Fennel (*Foeniculum vulgare*) is a flowering plant species in the celery family Apiaceae or Umbelliferae. It is treated as the sole species in the genus *Foeniculum* by most botanists. It is a hardy, perennial, herb, with yellow flowers and feathery leaves. It is indigenous to the shores of the Mediterranean but has become widely naturalized in many parts of the world, especially on dry soils near the sea-coast and on riverbanks. It is a highly aromatic and flavourful herb with culinary and medicinal uses and, along with the similar-tasting anise, is one of the primary ingredients of absinthe. Florence fennel or finocchio is a selection with a swollen, bulb-like stem base that is used as a vegetable.

Fennel is used as a food plant by the larvae of some Lepidoptera species including the mouse moth and the anise swallowtail.

Cultivation and Uses

Fennel is widely cultivated, both in its native range and elsewhere, for its edible, strongly flavoured leaves and fruits. Its aniseed flavour comes from anethole, an aromatic compound also found in anise and star anise, and its taste and aroma are similar to theirs, though usually not as strong.

The Florence fennel (*Foeniculum vulgare* Azoricum Group; syn. *F. vulgare* var. *azoricum*) is a cultivar group with inflated leaf bases which form a bulb-like structure. It is of cultivated origin, and has a mild anise-like flavour, but is more aromatic and sweeter. Florence fennel plants are smaller than the wild type. Their inflated leaf bases are eaten as a vegetable, both raw and cooked. There are several cultivars of Florence fennel, which is also known by several other names, notably the Italian name *finocchio*. In North American supermarkets, it is often mislabelled as "anise".

Foeniculum vulgare 'Purpureum' or 'Nigra', "bronze-leaved" fennel, is widely available as a decorative garden plant.

Fennel has become naturalised along roadsides, in pastures, and in other open sites in many regions, including northern Europe, the United States, southern Canada and in much of Asia and Australia. It propagates well by seed, and is considered an invasive species and a weed in Australia and the United States. In western North America, fennel can be found from the coastal and inland wildland-urban interface east into hill and mountain areas, excluding desert habitats.

Florence fennel is one of the three main herbs used in the preparation of absinthe, an alcoholic mixture which originated as a medicinal elixir in Switzerland and became, by the late 19th century, a popular alcoholic drink in France and other countries.

Culinary Uses

The bulb, foliage, and seeds of the fennel plant are widely used in many of the culinary traditions of the world. The small flowers of wild fennel (mistakenly known in America as fennel "pollen") are the most potent form of fennel, but also the most expensive. Dried fennel seed is an aromatic, anise-flavoured spice, brown or green in colour when fresh, slowly turning a dull grey as the seed ages. For cooking, green seeds are optimal. The leaves are delicately flavoured and similar in shape to those of dill. The bulb is a crisp vegetable that can be sautéed, stewed, braised, grilled, or eaten raw. They are used for garnishes and to add flavour to salads. They are also added to sauces and served with pudding. The leaves used in soups and fish sauce and sometimes eaten raw as salad.

Fennel seeds are sometimes confused with those of anise, which are similar in taste and appearance, though smaller. Fennel is also used as a flavouring in some natural toothpastes. The seeds are used in cookery and sweet desserts.

Many cultures in India, Pakistan, Afghanistan, Iran and the Middle East use fennel seed in their cookery. It is one of the most important spices in Kashmiri Pandit and Gujarati cooking. It is an essential ingredient of the Assamese/Bengali/Oriya spice mixture *panch phoron* and in Chinese five-spice powders. In many parts of India and Pakistan, roasted fennel seeds are consumed as *mukhwas*, an after-meal digestive and breath freshener. Fennel leaves are used as leafy green vegetables either by themselves or mixed with other vegetables, cooked to be served and consumed as part of a meal, in some parts of India. In Syria and Lebanon, it is used to make a special kind of egg omelette (along with onions, and flour) called *ijjeh*.

Many egg, fish, and other dishes employ fresh or dried fennel leaves. Florence fennel is a key ingredient in some Italian and German salads, often tossed with chicory and avocado, or it can be braised and served as a warm side dish. It may be blanched or marinated, or cooked in risotto.

Fennel seeds are the primary flavour component in Italian sausage.

In Spain the stems of the fennel plant are used in the preparation of pickled eggplants, "berenjenas de Almagro".

An herbal tea or tisane can be made from fennel.

Tomato a Fruit Vegetable

The tomato is the edible, often red fruit/berry of the nightshade *Solanum lycopersicum*, commonly known as a tomato plant. The species originated in the South American Andes and its use as a food originated in Mexico, and spread throughout the world following the Spanish colonization of the Americas. Its many varieties are now widely grown, sometimes in greenhouses in cooler climates.

The tomato is consumed in diverse ways, including raw, as an ingredient in many dishes, sauces, salads, and drinks. While it is botanically a fruit, it is considered a vegetable for culinary purposes (as well as under U.S. customs regulations), which has caused some confusion. The fruit is rich in lycopene, which may have beneficial health effects.

The tomato belongs to the nightshade family, *Solanaceae*. The plants typically grow to 1–3 metres (3–10 ft) in height and have a weak stem that often sprawls over the ground and vines over other plants. It is a perennial in its native habitat, although often grown outdoors in temperate climates as an annual.

History

Etymology: The word "tomato" comes from the Spanish *tomate*, which in turn comes from the Nahuatl word *tomatotl* [aĐÈtomatlal]. It first appeared in print in 1595. A member of the deadly nightshade family, tomatoes were erroneously thought to be poisonous (although the leaves are) by Europeans who were suspicious of their bright, shiny fruit. Native versions were small, like cherry tomatoes, and most likely yellow rather than red. The tomato is native to western South America and Central America.

Mesoamerica

Aztecs and other peoples in Mesoamerica used the fruit in their cooking. The exact date of domestication is unknown: by 500 BC, it was already being cultivated in southern Mexico and probably other areas. The Pueblo people are thought to have believed that those who witnessed the ingestion of tomato seeds were blessed with powers of divination. The large, lumpy tomato, a mutation from a smoother, smaller fruit, originated in Mesoamerica, and may be the direct ancestor of some modern cultivated tomatoes.

Spanish Distribution

Spanish conquistador Hernán Cortés may have been the first to transfer the small yellow tomato to Europe after he captured the Aztec city of Tenochtitlan, now Mexico City, in 1521, although Christopher Columbus, a Genoese working for the Spanish monarchy, may have taken them back as early as 1493. The earliest discussion of the tomato in European literature appeared in an herbal written in 1544 by Pietro Andrea Mattioli, an Italian physician and botanist, who suggested that a new type of eggplant had been brought to Italy that was blood red or golden colour when mature and could be divided into segments and eaten like an eggplant—that is, cooked and seasoned with salt, black pepper, and oil. However it wasn't until ten years later that tomatoes were named in print by Mattioli as *pomi d'oro*, or "golden apple".

After the Spanish colonization of the Americas, the Spanish distributed the tomato throughout their colonies in the Caribbean. They also took it to the Philippines, from where it spread to southeast Asia and then the entire Asian continent. The Spanish also brought the tomato to Europe. It grew easily in Mediterranean climates, and cultivation began in the 1540s.

It was probably eaten shortly after it was introduced, and was certainly being used as food by the early 17th century in Spain. The earliest discovered cookbook with tomato recipes was published in Naples in 1692, though the author had apparently obtained these recipes from Spanish sources. In certain areas of Italy, such as Florence, however, the fruit was used solely as a tabletop decoration before it was incorporated into the local cuisine in the late 17th or early 18th century.

Italy

The recorded history of tomatoes in Italy dates back to 31 October 1548 when the house steward of Cosimo de' Medici, the grand duke of Tuscany, wrote to the Medici private secretary informing him that the basket of tomatoes sent from the grand duke's Florentine estate at Torre del Gallo "had arrived safely." Tomatoes were grown mainly as ornamentals early on after their arrival in Italy.

For example, the Florentine aristocrat Giovanvettorio Soderini wrote how they "were to be sought only for their beauty" and were grown only in gardens or flower beds. The tomato's ability to mutate and create new and different varieties helped contribute to its success and spread throughout Italy. However, even in areas where the climate supported growing tomatoes, their proximity of growing to the ground suggested low status. They were not adopted as a staple of the peasant population because they were not as filling as other fruits already available. Additionally, both toxic and inedible varieties discouraged many people from attempting to consume or prepare them.

Unique varieties were developed over the next several hundred years for uses such as dried tomatoes, sauce tomatoes, pizza tomatoes, and tomatoes for long term storage. These varieties are usually known for their place of origin as much as by a variety name.

For example, Pomodorino del Piennolo del Vesuvio is the "hanging tomato of Vesuvius". Five different varieties have traditionally been used to make these "hanging" tomatoes. They are Fiaschella, Lampadina, Patanara, Principe Borghese, and Re Umberto. Other tomatoes that originated in Italy include San Marzano, Borgo Cellano, Christopher Columbus, Costoluto Genovese, and Italian Pear. These tomatoes are characterized by relatively intense flavour compared to varieties typically grown elsewhere.

Britain

Tomatoes were not grown in England until the 1590s. One of the earliest cultivators was John Gerard, a barber-surgeon. Gerard's *Herbal*, published in 1597, and largely plagiarized from continental sources, is also one of the earliest discussions of the tomato in England. Gerard knew the tomato was eaten in Spain and Italy. Nonetheless, he believed it was poisonous (in fact, the plant and raw fruit do have low levels of tomatine, but are not generally dangerous). Gerard's views were influential, and the tomato was considered unfit for eating (though not necessarily poisonous) for many years in Britain and its North American colonies. By the mid-18th century, tomatoes were widely eaten in Britain, and before the end of that century, the *Encyclopædia Britannica* stated the tomato was "in daily use" in soups, broths, and as a garnish. They were not part of the average man's diet, however, and though by 1820 they were described as "to be seen in great abundance in all our vegetable markets" and to be "used by all our best cooks", reference was made to their cultivation in gardens still "for the singularity of their appearance", while their use in cooking was associated with Italian or Jewish cuisine.

Middle East and North Africa

The tomato was introduced to cultivation in the Middle East by John Barker, British consul in Aleppo *circa* 1799 to 1825. Nineteenth century descriptions of its consumption are uniformly as an ingredient in a cooked dish. In 1881, it is described as only eaten in the region "within the last forty years". Today, the tomato is a critical and ubiquitous part of Middle Eastern cuisine, served fresh in salads (e.g. Arab salad, Israeli salad, Shirazi salad and Turkish salad), grilled with kebabs and other dishes, made into sauces, and so on. The early name used for tomato in Iran was *Armani badenjan* (Armenian eggplant). Currently, the name used for tomato in Iran is *gojeh farangi* [exotic plum].

North America

The earliest reference to tomatoes being grown in British North America is from 1710, when herbalist William Salmon reported seeing them in what is today South Carolina. They may have been introduced from the Caribbean. By the mid-18th century, they were cultivated on some Carolina plantations, and probably in other parts of the Southeast as well. Possibly, some people continued to think tomatoes were poisonous at this time; and in general, they were grown more as ornamental plants than as food. Thomas Jefferson, who ate tomatoes in Paris, sent some seeds back to America.

Alexander W. Livingston was the first person who succeeded in upgrading the wild tomato, developing different breeds and stabilizing the plants. In the 1937 yearbook of the Federal Department of Agriculture, it was declared that "half of the major varieties were a result of the abilities of the Livingstons to evaluate and perpetuate superior material in the tomato". Livingston's first breed of tomato, the Paragon, was introduced in 1870. In 1875, he introduced the Acme, which was said to be involved in the parentage of most of the tomatoes introduced by him and his competitors for the next twenty-five years.

When Alexander W. Livingston had begun his attempts to develop the tomato as a commercial crop, his aim had been to grow tomatoes smooth in contour, uniform in size and having better flavour. One year, after many attempts, he passed through his fields, picking out particular tomato plants having distinct characteristics and heavy foliage. He saved the seeds carefully. The following spring he set two rows across his family garden located just below the hill and milk house. To his happy surprise, each plant bore perfect tomatoes like the parent vine. After five years, the fruit became fleshier and larger. In 1870, Alexander introduced the Paragon and tomato culture soon became a great enterprise in the county. Today, the crop is grown in every state in the Union. He eventually developed over seventeen different varieties of the tomato plant.

Because of the long growing season needed for this heat-loving crop, several states in the US Sun Belt became major tomato-producers, particularly Florida and California. In California, tomatoes are grown under irrigation for both the fresh fruit market and for canning and processing. The University of California, Davis (UC Davis) became a major centre for research on the tomato. The C.M. Rick Tomato Genetics Resource Centre at UC Davis is a gene bank of wild relatives, monogenic mutants and miscellaneous genetic stocks of tomato. The Centre is named for the late Dr. Charles M. Rick, a pioneer in tomato genetics research. Research on processing tomatoes is also conducted by the California Tomato Research Institute in Escalon, California.

In California, growers have used a method of cultivation called dry-farming, especially with Early Girl tomatoes. This technique encourages the plant to send roots deep to find existing moisture in soil that retains moisture, such as clayey soil.

Modern Commercial Varieties

The poor taste and lack of sugar in modern garden and commercial tomato varieties resulted from breeding tomatoes to ripen uniformly red. This change occurred after discovery of a variety in the mid 20th century that ripened uniformly. This was widely cross-bred to produce red fruit without the typical green ring around the stem on uncross-bred varieties. Prior to general introduction of this trait, most tomatoes produced more sugar during ripening, and were sweeter and more flavourful.

***Figure:** Tomatoes that have not ripened uniformly*

Cultivation

The tomato is now grown worldwide for its edible fruits, with thousands of cultivars having been selected with varying fruit types, and for optimum growth in differing growing conditions. Cultivated tomatoes vary in size, from tomberries, about 5 mm in diametre, through cherry tomatoes, about the same 1–2 cm (0.4–0.8 in) size as the wild tomato, up to beefsteak tomatoes 10 cm (4 in) or more in diametre. The most widely grown commercial tomatoes tend to be in the 5–6 cm (2.0–2.4 in) diametre range. Most cultivars produce red fruit, but a number of cultivars with yellow, orange, pink, purple, green, black, or white fruit are also available. Multicoloured and striped fruit can also be quite striking. Tomatoes grown for canning and sauces are often elongated, 7–9 cm (3–4 in) long and 4–5 cm (1.6–2.0 in) diametre; they are known as plum tomatoes, and have a lower water content. Roma-type tomatoes are important cultivars in the Sacramento Valley.

Tomatoes are one of the most common garden fruits in the United States and, along with zucchini, have a reputation for outproducing the needs of the grower.

Quite a few seed merchants and banks provide a large selection of heirloom seeds. The definition of an heirloom tomato is vague, but unlike commercial hybrids, all are self-pollinators that have bred true for 40 years or more.

About 150 million tons of tomatoes were produced in the world in 2009. China, the largest producer, accounted for about one quarter of the global output, followed by United States and India. For one variety, plum or processing tomatoes, California accounts for 90% of U.S. production and 35% of world production.

Within the EU, there are several areas that grow tomatoes with Protected Geographical Status. These include:

- Pomodoro di Pachino (PGI), in Sicily
- Pomodoro S. Marzano dell'Agro Sarnese-Nocerino (PDO), in south Italy
- Tomaten von der Insel Reichenau (PGI), from Reichenau Island, Germany
- Pomodorino del Piennolo del Vesuvio (PDO), in Mt Vesuvius area.

Varieties

There are around 7500 tomato varieties grown for various purposes. Heirloom tomatoes are becoming increasingly popular, particularly among home gardeners and organic producers, since they tend to produce more interesting and flavourful crops at the cost of disease resistance and productivity. In 1973, Israeli scientists developed the world's first long shelf-life commercial tomato varieties.

Hybrid plants remain common, since they tend to be heavier producers, and sometimes combine unusual characteristics of heirloom tomatoes with the ruggedness of conventional commercial tomatoes.

Tomato varieties are roughly divided into several categories, based mostly on shape and size.

- "Slicing" or "globe" tomatoes are the usual tomatoes of commerce, used for a wide variety of processing and fresh eating.
- Beefsteak tomatoes are large tomatoes often used for sandwiches and similar applications. Their kidney-bean shape, thinner skin, and shorter shelf life makes commercial use impractical.
- Oxheart tomatoes can range in size up to beefsteaks, and are shaped like large strawberries.
- Plum tomatoes, or paste tomatoes (including pear tomatoes), are bred with a higher solids content for use in tomato sauce and paste, and are usually oblong.
- Pear tomatoes are pear-shaped, and are based upon the San Marzano types for a richer gourmet paste.
- Cherry tomatoes are small and round, often sweet tomatoes generally eaten whole in salads.

- Grape tomatoes, a more recent introduction, are smaller and oblong, a variation on plum tomatoes, and used in salads.
- Campari tomatoes are also sweet and noted for their juiciness, low acidity, and lack of mealiness. They are bigger than cherry tomatoes, but are smaller than plum tomatoes.

Early tomatoes and cool-summer tomatoes bear fruit even where nights are cool, which usually discourages fruit set. There are also varieties high in beta carotenes and vitamin A, hollow tomatoes and tomatoes that keep for months in storage.

Tomatoes are also commonly classified as determinate or indeterminate. Determinate, or bush, types bear a full crop all at once and top off at a specific height; they are often good choices for container growing. Determinate types are preferred by commercial growers who wish to harvest a whole field at one time, or home growers interested in canning. Indeterminate varieties develop into vines that never top off and continue producing until killed by frost. They are preferred by home growers and local-market farmers who want ripe fruit throughout the season. As an intermediate form, there are plants sometimes known as vigorous determinate or semideterminate; these top off like determinates, but produce a second crop after the initial crop. The majority of heirloom tomatoes are indeterminate, although some determinate heirlooms exist.

Most modern tomato cultivars are smooth surfaced, but some older tomato cultivars and most modern beefsteaks often show pronounced ribbing, a feature that may have been common to virtually all pre-Columbian cultivars. While virtually all commercial tomato varieties are red, some cultivars – especially heirlooms – produce fruit in other colours, including green, yellow, orange, pink, black, brown, ivory, white, and purple. Such fruits are not widely available in grocery stores, nor are their seedlings available in typical nurseries, but they can be bought as seed. Less common variations include fruit with stripes (Green Zebra), fuzzy skin on the fruit (Fuzzy Peach, Red Boar), multiple colours (Hillbilly, Burracker's Favourite, Lucky Cross), etc.

There is also a considerable gap between commercial and home-gardener cultivars. Home cultivars are often bred for flavour to the exclusion of all other qualities, while commercial cultivars are bred for factors like consistent size and shape, disease and pest resistance, suitability for mechanized picking and shipping, and ability to ripen after picking.

Tomatoes grow well with seven hours of sunlight a day. A fertilizer with an NPK ratio of 5-10-10 is often sold as tomato fertilizer or vegetable fertilizer, although manure and compost are also used.

Diseases and Pests

Tomato cultivars vary widely in their resistance to disease. Modern hybrids focus on improving disease resistance over the heirloom plants. One common tomato disease is tobacco mosaic virus. Handling cigarettes and other infected tobacco products can transmit the virus to tomato plants. Various forms of mildew and blight are also common tomato afflictions, which is why tomato cultivars are often marked with a combination of letters that refer to specific disease resistance. The most common letters are: V – *verticillium* wilt, F – *fusarium* wilt strain I, FF – *fusarium* wilt strain I and II, N – *nematodes*, T – *tobacco mosaic virus*, and A – *alternaria.*

Another particularly dreaded disease is curly top, carried by the beet leafhopper, which interrupts the lifecycle, ruining a nightshade plant as a crop. As the name implies, it has the symptom of making the top leaves of the plant wrinkle up and grow abnormally.

Some common tomato pests are stink bugs, cutworms, tomato hornworms and tobacco hornworms, aphids, cabbage loopers, whiteflies, tomato fruitworms, flea beetles, red spider mite, slugs, and Colorado potato beetles.

Tomato plants produce the plant peptide hormone systemin after an insect attack. Systemin activates defensive mechanisms, such as the production of protease inhibitors to slow the growth of insects. The hormone was first identified in tomatoes, but similar proteins have been identified in other species since.

Companion Plants

Tomatoes serve, or are served by, a large variety of companion plants. In fact, one of the most famous pairings is the tomato plant and carrots; studies supporting this relationship having produced a popular book about companion planting, *Carrots Love Tomatoes.*

Additionally, the devastating tomato hornworm has a major predator in various parasitic wasps, whose larvae devour the hornworm, but whose adult form drinks nectar from tiny-flowered plants like umbellifers. Several species of umbellifer are therefore often grown with tomato plants, including parsley, queen anne's lace, and occasionally dill. These also attract predatory flies that attack various tomato pests.

On the other hand, borage is thought to actually repel the tomato hornworm moth.

Other plants with strong scents, like alliums (onions, chives, garlic) and mints (basil, oregano, spearmint) are simply thought to mask the scent of the tomato plant, making it harder for pests to locate it, or to provide an alternative landing point, reducing the odds of the pests from attacking the correct plant. These plants may also subtly impact the flavour of tomato fruit.

Ground cover plants, including mints, also stabilize moisture loss around tomato plants and other solaneae, which come from very humid climates, and therefore may prevent moisture-related problems like blossom end rot. Finally, tap-root plants like dandelions break up dense soil and bring nutrients from down below a tomato plant's reach, possibly benefiting their companion.

Tomato plants, on the other hand, protect asparagus from asparagus beetles, because they contain solanum that kills this pest, while asparagus plants (as well as marigolds) contain a chemical that repels root nematodes known to attack tomato plants.

Pollination

Figure: *Tomato flower in full bloom, associated with a young, developing fruit.*

Figure: *The flower and leaves are visible in this photo of a tomato plant.*

In the wild, original state, tomatoes required cross-pollination; they were much more self-incompatible than domestic cultivars. As a floral device to reduce selfing, the pistil of wild tomatoes extends farther out of the flower than today's cultivars.

The stamens were, and remain, entirely within the closed corolla.

As tomatoes were moved from their native areas, their traditional pollinators, (probably a species of halictid bee) did not move with them. The trait of self-fertility became an advantage, and domestic cultivars of tomato have been selected to maximize this trait.

This is not the same as self-pollination, despite the common claim that tomatoes do so. That tomatoes pollinate themselves poorly without outside aid is clearly shown in greenhouse situations, where pollination must be aided by artificial wind, vibration of the plants (one brand of vibrator is a wand called an "electric bee" that is used manually), or more often today, by cultured bumblebees.

The anther of a tomato flower is shaped like a hollow tube, with the pollen produced within the structure, rather than on the surface, as in most species. The pollen moves through pores in the anther, but very little pollen is shed without some kind of outside motion.

The best source of outside motion is a sonicating bee, such as a bumblebee, or the original wild halictid pollinator. In an outside setting, wind or animals provide sufficient motion to produce commercially viable crops.

Hydroponic and Greenhouse Cultivation

Tomatoes are often grown in greenhouses in cooler climates, and there are cultivars such as the British 'Moneymaker' and a number of cultivars grown in Siberia that are specifically bred for indoor growing. In more temperate climates, it is not uncommon to start seeds in greenhouses during the late winter for future transplant.

Greenhouse tomato production in large-acreage commercial greenhouses and owner-operator stand-alone or multiple-bay greenhouses is on the increase, providing fruit during those times of the year when field-grown fruit is not readily available.

Smaller sized fruit (cherry and grape), or cluster tomatoes (fruit-on-the-vine) are the fruit of choice for the large commercial greenhouse operators while the beefsteak varieties are the choice of owner-operator growers.

Hydroponic tomatoes are also available, and the technique is often used in hostile growing environments, as well as high-density plantings.

Picking and Ripening

To facilitate transportation and storage, tomatoes are often picked unripe (green) and ripened in storage with ethylene. Unripe tomatoes are firm. As they ripen they soften until reaching the ripe state where they are red or orange in colour and slightly soft to the touch. Ethylene is a hydrocarbon gas produced by many fruits that acts as the molecular cue to begin the ripening process. Tomatoes ripened in this way tend to keep longer, but have poorer flavour and a mealier, starchier texture than tomatoes ripened on the plant. They may be recognized by their colour, which is more pink or orange than the other ripe tomatoes' deep red, depending on variety.

A machine-harvestable variety of tomato (the "square tomato") was developed in the 1950s by University of California, Davis's Gordie C. Hanna, which, in combination with the development of a suitable harvester, revolutionized the tomato-growing industry. This type of tomato is grown commercially near plants that process and can tomatoes, tomato sauce, and tomato paste. They are harvested when ripe and are flavourful when picked. They are harvested 24 hours a day, seven days a week during a 12 to 14 week season, and immediately transported to packing plants, which operate on the same schedule. California is a centre of this sort of commercial tomato production and produces about a third of the processed tomatoes produced in the world.

In 1994, Calgene introduced a genetically modified tomato called the *FlavrSavr*, which could be vine ripened without compromising shelf life. However, the product was not commercially successful, and was sold only until 1997. Slow-ripening cultivars of tomato have been developed by crossing a non-ripening cultivar with ordinary cultivars. Cultivars were selected whose fruits have a long shelf life and at least reasonable flavour.

At home, fully ripe tomatoes can be stored in the refrigerator, but are best kept at room temperature. Tomatoes stored cold remain edible, but tend to lose their flavour permanently. Tomatoes stored stem down may also keep from rotting too quickly.

Genetic Modification

Tomatoes that have been modified using genetic engineering have been developed, and although none are commercially available now, they have been in the past. The first commercially available genetically modified food was a variety of tomato named (the Flavr Savr), which was engineered to have a longer shelf life. Scientists are continuing to develop tomatoes with new traits not found in natural crops, such as increased resistance to pests or environmental stresses. Other projects aim to enrich tomatoes with substances that may offer health benefits or provide better nutrition.

Consumption

The tomato is now grown and eaten around the world. It is used in diverse ways, including raw in salads, and processed into ketchup or tomato soup. Unripe green tomatoes can also be breaded and fried, used to make salsa, or pickled. Tomato juice is sold as a drink, and is used in cocktails such as the Bloody Mary.

Tomatoes are acidic, making them especially easy to preserve in home canning whole, in pieces, as tomato sauce or paste. The fruit is also preserved by drying, often in the sun, and sold either in bags or in jars with oil.

Tomatoes are used extensively in Mediterranean cuisine. They are a key ingredient in pizza, and are commonly used in pasta sauces. They are also used in gazpacho (Spanish cuisine) and *pa amb tomàquet* (Catalan cuisine).

Though it is botanically a berry, a subset of fruit, the tomato is a vegetable for culinary purposes, because of its savory flavour.

Nutrition

Tomatoes are now eaten freely throughout the world. They contain the carotene lycopene, one of the most powerful natural antioxidants. In some studies, lycopene, especially in cooked tomatoes, has been found to help prevent prostate cancer, but other research contradicts this claim. Lycopene has also been shown to improve the skin's ability to protect against harmful UV rays.

A study done by researchers at Manchester and Newcastle universities revealed that tomato can protect against sunburn and help keeping the skin looking youthful. Natural genetic variation in tomatoes and their wild relatives has given a genetic plethora of genes that produce lycopene, carotene, anthocyanin, and other antioxidants. Tomato varieties are available with double the normal vitamin C (Doublerich), 40 times normal vitamin A (97L97), high levels of anthocyanin (resulting in blue tomatoes), and two to four times the normal amount of lycopene (numerous available cultivars with the high crimson gene).

Potential Health Benefits

Lycopene has also been shown to protect against oxidative damage in many epidemiological and experimental studies. In addition to its antioxidant activity, other metabolic effects of lycopene have also been demonstrated. The richest source of lycopene in the diet is tomato and tomato derived products. Tomato consumption has been associated with decreased risk of breast cancer, head and neck cancers and might be strongly protective against neurodegenerative diseases.

Tomatoes, tomato sauces and puree are said to help lower urinary tract symptoms (BPH) and may have anticancer properties. Tomato consumption might be beneficial for reducing cardiovascular risk associated with type 2 diabetes.

Storage

Tomatoes keep best unwashed at room temperature and out of direct sunlight. It is not recommended to refrigerate as this can harm the flavour. Tomatoes that are not yet ripe can be kept in a paper bag till ripening. Storing stem down can prolong shelf life.

Safety

Plant Toxicity

Leaves, stems, and green unripe fruit of the tomato plant contain small amounts of the toxic alkaloid tomatine. They also contain solanine, a toxic alkaloid found in potato leaves and other plants in the nightshade family. Use of tomato leaves in herbal tea has been responsible for at least one death.

However, levels of tomatine in foliage and green fruit are generally too small to be dangerous unless large amounts are consumed, for example, as greens. Small amounts of tomato foliage are sometimes used for flavouring without ill effect, and the green fruit is sometimes used for cooking, particularly as fried green tomatoes. Compared to potatoes the amount of solanine in green or ripe tomatoes is low; however, even in the case of potatoes while solanine poisoning resulting from dosages several times normal human consumption has been demonstrated, actual cases of poisoning resulting from excessive consumption of potatoes that have high concentration of solanine are rare.

Tomato plants can be toxic to dogs if they eat large amounts of the fruit, or chew plant material.

Salmonella

On 30 October 2006, the U.S. Centres for Disease Control and Prevention (CDC) announced tomatoes might have been the source of a salmonellosis outbreak causing 172 illnesses in 18 states. Tomatoes have been linked to seven salmonella outbreaks since 1990.

The 2008 United States salmonellosis outbreak caused the removal of tomatoes from stores and restaurants across the United States and parts of Canada, although other foods, including jalapeño and serrano peppers, may have been involved.

Botanical Description

Tomato plants are vines, initially decumbent, typically growing six feet or more above the ground if supported, although erect bush varieties have been bred, generally three feet tall or shorter. Indeterminate types are "tender" perennials, dying annually in temperate climates (they are originally native to tropical highlands), although they can live up to three years in a greenhouse in some cases. Determinate types are annual in all climates.

Tomato plants are dicots, and grow as a series of branching stems, with a terminal bud at the tip that does the actual growing. When that tip eventually stops growing, whether because of pruning or flowering, lateral buds take over and grow into other, fully functional, vines.

Tomato vines are typically pubescent, meaning covered with fine short hairs. These hairs facilitate the vining process, turning into roots wherever the plant is in contact with the ground and moisture, especially if the vine's connection to its original root has been damaged or severed.

Most tomato plants have compound leaves, and are called regular leaf (RL) plants, but some cultivars have simple leaves known as potato leaf (PL) style because of their resemblance to that particular relative. Of RL plants, there are variations, such as rugose leaves, which are deeply grooved, and variegated, angora leaves, which have additional colours where a genetic mutation causes chlorophyll to be excluded from some portions of the leaves.

The leaves are 10–25 cm (4–10 in) long, odd pinnate, with five to 9 leaflets on petioles, each leaflet up to 8 cm (3 in) long, with a serrated margin; both the stem and leaves are densely glandular-hairy.

Their flowers, appearing on the apical meristem, have the anthers fused along the edges, forming a column surrounding the pistil's style. Flowers in domestic cultivars tend to be self-fertilizing. The flowers are 1–2 cm (0.4–0.8 in) across, yellow, with five pointed lobes on the corolla; they are borne in a cyme of three to 12 together.

Tomato fruit is classified as a berry. As a true fruit, it develops from the ovary of the plant after fertilization, its flesh comprising the pericarp walls. The fruit contains hollow spaces full of seeds and moisture, called locular cavities. These vary, among cultivated species, according to type. Some smaller varieties have two cavities, globe-shaped varieties typically have three to five, beefsteak tomatoes have a great number of smaller cavities, while paste tomatoes have very few, very small cavities.

For propagation, the seeds need to come from a mature fruit, and be dried or fermented before germination.

Botanical Classification

In 1753, Linnaeus placed the tomato in the genus *Solanum* (alongside the potato) as *Solanum lycopersicum*. In 1768, Philip Miller moved it to its own genus, naming it *Lycopersicon esculentum*. This name came into wide use, but was in breach of the plant naming rules. Technically, the name *Lycopersicon lycopersicum* would be more correct, but is rarely used (except in seed catalogues, which frequently used it and still do).

Genetic evidence has now shown that Linnaeus was correct to put the tomato in the genus *Solanum*, making *Solanum lycopersicum* the correct name. Both names, however, will probably be found in the literature for some time. Two of the major reasons some still consider the genera separate are the leaf structure (tomato leaves are markedly different from any other *Solanum*), and the biochemistry (many of the alkaloids common to other *Solanum* species are conspicuously absent in the tomato). Hybrids of tomato and diploid potato can be created in the lab by somatic fusion, and are partially fertile, providing evidence of the close relationship between these species.

Wild Species

Including Solanum lycopersicum, there are currently 13 species recognized in Solanum section Lycopersicon. Three of these species—S. Cheesmaniae, S. Galapagense, and S. Pimpinellifolium—are fully cross compatible with domestic tomato. Four more species—S. chmielewskii, S. habrochaites, S. neorickii, and S. pennelli—can be readily crossed with domestic tomato, with some limitations. Five species—S. arcanum, S. chilense, S. corneliomulleri, S. huaylasense, and S. peruvianum—can be crossed with domestic tomato with difficulty and usually require embryo rescue to produce viable plants. The Lycopersicon section has not been fully sampled within wild species in the South American range, so new species may be added in the future.

Solanum section Lycopersicoides and section Juglandifolium are represented by two species each that are considered bridge species genetically intermediate between tomato and non-tuber bearing potato species. S. Lycopersicoides can be crossed with domestic tomato and introgression lines have been developed. This species was significant in moving the domestic tomato from separate genus status into the Solanum group because it directly links the tomato into the potato family.

Genome Sequencing

An international consortium of researchers from 10 countries, among them researchers from the Boyce Thompson Institute for Plant Research, began sequencing the tomato genome in 2004, and is creating a database of genomic sequences and information on the tomato and related plants. A prerelease version of the genome was made available in December 2009. The genomes of its mitochondria and chloroplasts are also being sequenced as part of the project. The complete genome for the cultivar Heinz 1706 was published on 31 May 2012 in *Nature*. Since many other fruits, like strawberries, apples, melons, and bananas share the same characteristics and genes, researchers stated the published genome could help to improve food quality, food security and reduce costs of all of these fruits.

Breeding

Active breeding programmes are ongoing by individuals, universities, corporations, and organizations. The Tomato Genetic Resource Centre, Germplasm Resources Information Network, AVRDC, and numerous seed banks around the world store seed representing genetic variations of value to modern agriculture. These seed stocks are available for legitimate breeding and research efforts. While individual breeding efforts can produce useful results, the bulk of tomato breeding work is at universities and major agriculture-related corporations. These efforts have resulted in significant regionally adapted breeding lines and hybrids, such as the Mountain series from North Carolina. Corporations including Heinz, Monsanto, BHNSeed, Bejoseed, etc., have breeding programmes that attempt to improve production, size, shape, colour, flavour, disease tolerance, pest tolerance, nutritional value, and numerous other traits.

Fruit or Vegetable?

Botanically, a tomato is a fruit: the ovary, together with its seeds, of a flowering plant. However, the tomato has a much lower sugar content than other edible fruits, and is therefore not as sweet. Typically served as part of a salad or main course of a meal, rather than at dessert, it is considered a vegetable for most culinary uses. One exception is that tomatoes are treated as a fruit in home canning practices: they are acidic enough to process in a water bath rather than a pressure cooker as vegetables require. Tomatoes are not the only food source with this ambiguity: green beans, eggplants, cucumbers, and squashes of all kinds (such as zucchini and pumpkins) are all botanically fruits, yet cooked as vegetables.

This dispute has led to legal speculation in the United States. In 1887, U.S. tariff laws that imposed a duty on vegetables, but not on fruits, caused the tomato's status to become a matter of legal importance. The U.S. Supreme Court settled this controversy on 10 May 1893, by declaring that the tomato is a vegetable, based on the popular definition that classifies vegetables by use, that they are generally served with dinner and not dessert (*Nix v. Hedden* (149 U.S. 304)). The holding of this case applies only to the interpretation of the Tariff Act of 3 March 1883, and the court did not purport to reclassify the tomato for botanical or other purposes.

Tomatoes have been designated the state vegetable of New Jersey. Arkansas took both sides by declaring the *South Arkansas Vine Ripe Pink Tomato* both the state fruit and the state vegetable in the same law, citing both its culinary and botanical classifications. In 2009, the state of Ohio passed a law making the tomato the state's official fruit. Tomato juice has been the official beverage of Ohio since 1965. A.W. Livingston, of Reynoldsburg, Ohio, played a large part in popularizing the tomato in the late 19th century; his efforts are commemorated in Reynoldsburg with an annual Tomato Festival.

Names

The scientific species epithet *lycopersicum* means "wolf peach", and comes from German werewolf myths. These legends said that deadly nightshade was used by witches and sorcerers in potions to transform themselves into werewolves, so the tomato's similar, but much larger, fruit was called the "wolf peach" when it arrived in Europe.

The native Mexican tomatillo is *tomate* (in Nahuatl: *tomâtl*, meaning 'fat water' or 'fat thing'). When Aztecs started to cultivate the Andean fruit, bigger and red, they called the new species *xitomatl* (or *jitomates*) (pronounced [*ƒ*iÐÈtomatl]), ('plump thing with navel' or 'fat water with navel'). After their conquest of Tenochtitlan, Spaniards exported tomatoes (*jitomates*) to the rest of the world with the name *tomate*, so numerous languages use forms of the word "tomato" (*tomate*) to refer to the red tomato instead of the green tomatillo.

The Italian word, *pomodoro* (from *pomo d'oro* "apple of gold") was borrowed into Polish, and via Russian, into several other languages. Similarly, the now rare German term *Paradeisapfel* (for "apple of paradise") is still heard in the form *paradeiser* in the Bavarian and Austrian dialects, and was borrowed into modern Hungarian, Slovenian and Serbian.

Turnips and Rutabagas

Grow turnips and rutabagas to bring rich, earthy flavours to your kitchen. This guide includes information on types of turnips and rutabagas and how to plant these versatile fall vegetables, plus recipes for scrumptious turnip greens.

Winter Grains

Learn how to grow wheat, rye, oats and barley to build soil fertility and provide forage for your animals as well as whole grains and flour for your kitchen.

Winter Squash

Enjoy scrumptious winter squash, an excellent storage crop, in its many diverse types. Learn how to plant, grow, harvest and store a range of winter squash, including butternut, acorn and spaghetti squash.

Bibliography

Arguello D. S.: *Production, Post-harvest Management and Exportation of Tropical Fruits in Costa Rica*, Wageningen, Technical Centre for Agricultural and Rural Cooperation, 1992.

Arora, Chandralekha : *Advances in Horticulture Science Research*, Oxford Book Company, Delhi, 2009.

Barbeau G.: *Tropical Fruits in Nicaragua*, Managua, Nicaragua Ministerio de Desarrollo Agropecuario, Agraria, 1990.

Bhattacharjee, Supriya Kumar : *Amenity Horticulture, Biotechnology and Post-Harvest Technology*, Pointer, Delhi, 2006.

Chadha K. L. and Pareek O. P.: *Advances in Horticulture: Fruit Crops,* New Delhi, Malhotra Publishing House, 1993.

Crucefix D.: *Avocado Variety Selection for Export Development,* Roseau, CARDI, 1996.

Currah L. and Proctor F. J.: *Onions in Tropical Regions*, Kent, Natural Resources Institute, 1990.

Dhatt A. S. and Singh Z.: *Propagation and Rootstocks of Citrus*, New Delhi, Malhotra Publishers, 1993.

Doijode S. D.: *Seed Germination in Fruits*, New Delhi, Malhotra Publishers, 1993.

Featherly H. I.: *Taxonomic Terminology of the Higher Plants*, USA, Iowa State College Press, 1954.

Gibbs, D.R. : *Chemotaxonomy of Flowering Plants*, McGill-Queen's University Press, Montreal, 1974.

Glowinski Louis: *The Complete Book of Fruit Growing in Australia*, Australia, Lothian Publishing Company Pty. Ltd., 1991.

Gunjate R. T. and Tawde A. B.: *Propagation and Rootstocks of Mango*, New Delhi, Malhotra, 1993.

Harlan, Jack R.: *The Living Fields: Our Agricultural Heritage*, Cambridge: Cambridge University Press, 1998.

Herklots G. A. C.: *Vegetables in South East Asia*, London, George Allen & Unwin Ltd., 1972.

Jeffers P.: *Evaluation of Four Onion Varieties in Montserrat*, Plymouth, CARDI, 1992.

Jerram, M.R.K. : *A Text-Book on Forest Management : In Agriculture, Horticulture and Forestry*, Asiatic Pub, Delhi, 2006.

Jugale, V.B. : *Horticulture Economy of Maharashtra*, Shruti, Delhi, 2004.

Malins A.: *Postharvest Handling of Pineapple and Mango*, Port of Spain, Trinidad and Tobago, 1992.

Matsuoka H.: *Cultivation of Panicum Genetic Resources for Evaluation of Characteristics*, Tokyo, JICA, 1997.

Mitra S.: *Postharvest Physiology and Storage of Tropical and Subtropical Fruits*, Oxon, CABI, 1997.

Morton Julia F.: *Fruits of Warm Climates*, Miami, Julia F. Morton Publisher, 1987.

Oldham P.: *Cost of Production of Major Tree Crops in Dominica*, Roseau, Ministry of Agriculture, 1991.

Peter, K.V. : *Basics of Horticulture*, New India, Delhi, 2009.

Pilgrim R.: *Post Harvest Handling of Minor Exotics*, St. George's, Grenada, 1996.

Prasad, Surendra : *Agro's Dictionary of Horticulture*, Agrobios, Delhi, 2003.

Sharma, V.K. : *Advances in Horticulture : Strategies, Production, Plant Protection and Value Addition*, Deep and Deep, Delhi, 2004.

Singh H. P. and Chadha K. L.: *Genetic Resources of Citrus*, New Delhi, Malhotra Publishing House, 1993.

Singh, D.K. : *Elements of Horticulture*, Agrotech, Delhi, 2005.

Singh, H.P. : *Challenges of Climate Change - Indian Horticulture*, Westville Publishing House, Delhi, 2010.

Singh, S.P. : *Advances in Horticulture and Forestry*, Scientific, Delhi, 2001.

Thavarasook C.: *Proceedings of the Mungbean Meeting 90, Chiang Mai, Thailand,* Bangkok, Tropical Agricultural Research Centre, 1991.

Weiss, R.F. : *Herbal Medicine*, Beaconsfield Arcanum, Beaconsfield, 1991 .

Whitwell A.: *Dominica Orchard Crop Management and Research Project: Pest and Disease Management,* United Kingdom, Natural Resources Institute, 1991.

Index

K

L

M

N

O

P

R

S

T

V

W

❑❑❑